SpringerBriefs in Astronomy

For further volumes:
http://www.springer.com/series/10090

Claudio Vita-Finzi

Solar History

An Introduction

 Springer

Claudio Vita-Finzi
Department of Mineralogy
Natural History Museum
London
UK

ISSN 2191-9100   ISSN 2191-9119   (electronic)
ISBN 978-94-007-4294-9   ISBN 978-94-007-4295-6   (eBook)
DOI 10.1007/978-94-007-4295-6
Springer Dordrecht Heidelberg New York London

Library of Congress Control Number: 2012943359

Printed on acid-free paper

Springer is part of Springer Science+Business Media (www.springer.com)

For Raphy and Sam

Preface

Approximately 5,000,000,000 years from now the Sun will exhaust its supply of hydrogen and turn into a red star so bloated that it swallows the planet Mercury and so hot that it melts the Earth's surface. That much is generally agreed upon. But what the Sun will do next year, let alone the next decade, is not at all clear. It will not be much bigger or nearer to us than now, and its net output will probably not depart from the average of the last few decades by more than a few tenths of one percent, but by how many tenths, and will the contribution of ultraviolet rays increase or decrease, and in either case how rapid will the departure be?

The answers bear not only on climate change but also on human health and the many activities—including the rest of astronomy, radio and satellite communications, geodesy and Space travel—where small changes in the Sun can have a disproportionate impact. To tackle these questions we require a secure grasp of the Sun's entire evolution so that long-term trends as well as short-term fluctuations can be properly assessed.

But solar history needs no utilitarian justification. It bears on our planet's evolution, and that of its occupants; it reflects events throughout the solar system; it illuminates the fate of the billions of stars in the Milky Way and beyond; and it illustrates the noble achievement of those who are making sense of an incandescent ball of gas with a volume a million times the Earth's and distant enough for its light to take 8 min to reach us. D. W. Hamlyn once suggested that a history of philosophy, apart 'from the strictly historical sense....ought to provide a due sense of the complexity and many-sidedness' of its subject. That is what I had in mind when writing *Solar History*.

Solar history is extremely uneven. Documentary sources are available for only 0.00001 % of the Sun's existence. For the remaining 4.5 aeons we depend on analogy with other stars, computer modelling, data derived from meteorites, polar ice caps and tree rings, the fossil record, and other indirect guides to solar behaviour. In *The Sun—a User's Manual* (2008) I hinted at these matters but necessarily focused on the present day while claiming that a history of the Sun was beginning to fall into place. There had in fact been three magisterial multiple-author surveys of the solar past [1–3], but much new information has been skilfully (and expensively) secured in the last two decades which bears on parts of the narrative.

I am indebted to Penelope Vita-Finzi for encouragement, Ken Phillips and Ilya Usoskin for their sage comments on parts of the text, Dom Fortes for reading the whole thing, Michael Wood for compassion, many individuals and organisations—above all NASA—for data and images, and Petra van Steenbergen at Springer for support.

Fenstead End, Suffolk, May 2012   Claudio Vita-Finzi

References

1. White OR (ed) (1977) The solar output and its variation. Colorado Associated University Press, Boulder
2. Pepin RO, Eddy JA, Merrill RB (eds) (1980) The ancient Sun. Pergamon, New York
3. Sonett CP, Giampapa MS, Matthews MS (eds) (1991) The Sun in time. University of Arizona Press, Tucson

Contents

Chapter 1
Introduction

Abstract Solar History traces the Sun's evolution from the origins of the Solar System some 4,500,000 yr ago to recent months. In doing so it fuses the data of astrophysicists with those of observational astronomers, field scientists and historians. The subject matter thus spans a wide range of timescales, from the aeons (Gyr, 10^9 yr) of the solar modellers to the millions of years (Myr, 10^6 yr) in which much of solar history is read from meteorites, the millennia and centuries (10^5–10^2 yr) documented by ice cores, tree rings and sunspots, and the decades, years, days and seconds (10^1–10^{-9} yr) of telescopic and satellite observational astronomy.

Keywords Solar history · Solar model · Meteorite · Ice core · Tree ring · Sunspot · Observational astronomy · Neutrino

The observational history of the Sun began with the systematic recording of sunspots 400 years ago; we have to use indirect sources to document the remaining 4,499,600 years, and many of those sources involve meteorites, ice cores and sediments rather than the kind of data conventionally handled by astronomers and astrophysicists. Such terrestrial data may well help to uncover evidence for solar variation that will help astrophysicists understand the internal dynamics of the Sun [9].

We have to deal with evidence at widely varying degrees of precision. Our knowledge of events in the earliest 4.5 Gyr appears sketchy and that of last year abundant. This would be a crude way to apportion space in a narrative, as sketchiness is a misleading term when we are dealing with cosmic dimensions and abundance does not always breed understanding. It seems more informative for successive chapters to discuss solar history at the appropriate resolution ranging from Gyr to minutes and seconds. As a crude analogy, consider the benefits that have come from juxtaposing the scanty evidence for the first few million years of human evolution with the physiology and behaviour of living individuals.

C. Vita-Finzi, *Solar History*, SpringerBriefs in Astronomy,
DOI: 10.1007/978-94-007-4295-6_1, © The Author(s) 2013

Sun and Earth

Changes in the Sun's appearance and apparent motion have doubtless been noted from deepest prehistoric times and used to underpin superstition and authority. The earliest record of a solar eclipse (in China) dates from the third millennium BC and its prediction (by the Babylonians) from the fourth century BC, but the first solar observations that qualify as scientific because they combine observation with some attempt at rational explanation were by early Greek astronomers, notably Aristarchus, who introduced a heliocentric model of the solar system in the third century BC and estimated the size of the Sun and Moon and their distances from the Earth.

In the fifth century BC Anaxagoras had argued that the Sun was a hot mass of metal. On the basis of experiment with spheres made of Earth-like materials the Count de Buffon (1707–1788) calculated that it would take the Earth 74,832 years to cool from white heat to its present temperature. Two other mechanisms for solar luminosity later gained support, a chemical process akin to combustion, and the energy liberated by gravity. A fourth option was the infall of meteorites, whose kinetic energy would be transformed into heat energy, but an endless supply of meteoroids within gravitational reach of the Sun was difficult to envisage.

The gravitational force exerted by the collapse of the Sun's solid constituents was central to the model proposed by Hermann von Helmholtz in 1854. It gave a maximum age of 20 million years. In 1887 William Thomson—Lord Kelvin—developed Helmholtz's theory by postulating that the Sun originated in a cloud of dust and gas (Fig. 1.1) which got hotter as it collapsed to the present dimensions of the Sun. The internal temperature of millions of degrees was prolonged by gravitational contraction: at a rate of 50 m/100 yr the Sun would remain hot for 100 million years.

On the more extreme assumption that the Sun started out molten, Kelvin ascribed to him (as he referred to the Sun) an age of 10–100 million years. In an article published in *Macmillan's Magazine* [20] he argued that, as the solar composition was much like the Earth's (sodium, iron, manganese and other metals had been discovered in its atmosphere), its specific heat is not very different from that of water, and, dividing its rate of radiation by its mass he obtained a value of 1.4 °C/yr, whereupon its diameter would fall by 1 % in 860 years. But that would surely have been noticed by astronomers, and in any case the contraction would have generated a similar amount of heat by 'mutual gravitation'. After further calculation and cogitation he concluded, if such ballpark figures can be termed conclusions, that the Sun's specific heat was 10–10,000 times that of liquid water and its temperature had therefore fallen 100 °C in 700–700,000 years.

Kelvin tested some of his ideas against geology. A solar contraction of 1 °C ran counter to geological evidence, and Charles Darwin's talk of 300 million years for the denudation of the Weald ran counter to Kelvin's calculations based on the extent to which the physical conditions on the Sun differed from laboratory data. And, scoff though he might at Darwin's estimates for the time required to

Fig. 1.1 Birth of a star. Artist's conception based on observation of L1014 by Spitzer Space Telescope. (Courtesy of NASA/JPLCaltech/R.Hurt (SSC))

accomplish erosion of the Weald of SE England (namely 300 million years), Kelvin's own claim that the Sun's surface temperature was not incomparably higher than temperatures obtained in the lab or in a locomotive furnace would be considered reckless if he had not added the qualification *unless sources now unknown to us are prepared in the great storehouse of creation.*

The phrase became famous when quoted by Ernest Rutherford, who had noticed Kelvin in the audience when he began a lecture on the newly-discovered phenomenon of radioactivity and its potential as a source of heat within the Earth. The source in question is of course not the same. For the Earth it is Rutherford's heat of radioactive decay; for the Sun it is nuclear fusion. But before the discovery of radioactivity geology in the broad sense seemed to support Kelvin. For example, G. H. Darwin showed that it would take an initially molten Earth 56 million years to acquire by tidal friction its present 24 h day [3], and even Huxley [12] argued that it was perhaps up to biology to make its data fit Kelvin's timescale rather than the other way round.

In any case Kelvin's calculations greatly underestimated the Earth's age not because he did not include a radioactivity factor but because he did not allow for convection in the Earth [6]. Rutherford pointed to radium, which has too short a half-life to do the trick, yet even if all the potential radioactive sources are included the heat yield is so low (a few mW/m^2 during the first Gyr of the Earth's history) that Kelvin's calculations are practically unaffected [18]. On the other hand, as John Perry [17] noted, convection in the mantle would mean that instead of the shallow outer layer of a solid sphere there would be available a much larger volume of hot rock to account for the present heat flux at the surface and thus a cooling history expressed in billions rather than millions of years.

Earth history still serves as a check on solar history by providing a minimum age for its creation, illuminating the role of impacts in solar system evolution, and revealing the effect of the solar wind —the charged particles ejected by the Sun's upper atmosphere—and solar flares on the Moon or on meteorites that originated in the protoplanetary disk. These roles are increasingly overshadowed by the activity of neutrino and helioseismological observatories and the findings of dedicated artificial satellites but they remain crucial to any attempt at tracing solar history.

Cycles and Trends

The classical historian Arnaldo Momigliano once commented that *Historians are supposed to be discoverers of truths. No doubt they must turn their research into some sort of story before being called historians. But their stories must be true stories. [...] History is no epic, history is no novel, history is not propaganda because in these literary genres control of the evidence is optional, not compulsory* [16, p. 265]. It is odd to see a classical historian preaching truth who is condemned to rely on forgeries, mistranslations and barefaced lies for much of his raw material, but the ambiguity of 'supposed' (presumably intended and not itself a victim of translation) breathes irony. At all events the present account does its best to safeguard the evidence.

That is not easy in view of the emphasis in the solar sources on cyclical behaviour. Consider for instance the solar cycle. The term is usually applied to the ~ 11 year periodicity by which sunspot numbers rise and fall and the spots migrate towards the Sun's lower latitudes before the process begins anew. The first statistical study, which identified the cycle, was by S. H. Schwabe in 1843, and much effort was expended thereafter in the search for a link between sunspot number and mean global temperature. But it was cycle *length*, sometimes presented as a statistical annoyance, which provided the first persuasive link between solar variation and climate [7]. The link has been found wanting [2] but cyclicity remains at the heart of much solar-terrestrial research.

Again, a secular, upward trend in irradiance emerged from satellite measurements after data from several instruments had been painstakingly stitched together [22]. Though the stitching, and thus the trend, have been challenged by other workers [8], the possibility of a trend attracts less comment than the quality of the cyclic records. That may be because, when the topic of conversation is global warming, discussion of the solar factor is done in a whisper for fear of attracting the label of climate change denier. Yet no analysis of solar physics can ignore the slightest hint of cumulative change even if in the event it proves to be mistaken.

Of course the satellite data, and indeed the historical sunspot records, are too short for any statistical (or even ocular) test to convince the sceptic and, more important, to underpin a physical model. The controversial satellite measurements run from 1978, when NASA's NIMBUS 7-ERB (Earth Radiation Budget) instrument was launched. At the time of writing that totals 33 years. Reliable observation of sunspots goes back a little over 400 years if we extend the sequence back to the first observations by Galileo and Thomas Harriot. That is less than 0.00001 % of the Sun's history.

The sunspot record has been extended back to AD 850, that is to say for an additional 750 years, by reference to the cosmogenic isotope beryllium-10 (^{10}Be) recovered from ice cores [21], and analysis of carbon-14 (^{14}C) from tree rings has prolonged the sunspot sequence to a total of 11,400 years [19]. Variations in the sunspot number are then used to compare levels of solar activity at different times. The agreement between calculated and observed sunspot numbers since 1610 is impressive; even so one may wonder whether the ^{14}C evidence could be used in a less roundabout way as a measure of solar activity. At all events, if sunspot activity is the target, 99.99999 % of the Sun's lifetime remains undocumented despite the addition of synthetic data.

The Need for a Long Chronology

Earth history was transformed by recognition of its length. To begin with, evidence that was otherwise inexplicable now became open to rational discourse, notably organic evolution and the products of observable processes of erosion and deposition. Then the door was open to processes which had yet to be imagined, such as major glaciations. And finally the Earth was seen to be an ephemeral constituent of a turbulent universe.

An appropriately long solar history for the Sun shares all the above but it shifts attention from processes on a human timescale and with human significance (such as sunspots and climate change) to stellar concerns. The variability of other suns, their atmospheric past and their meteoritic bruising are now issues whose study have become an integral part of the subject just as planetary exploration enriched Earth science.

The job has just begun. As Fig. 1.1 shows, we have information on solar events back to the time of the Sun's creation, but it is dominated by solar flares and the information embodied in crashed meteorites. This is not much worse than our grasp of events on the young Earth. In fact, it is better, as solar history can benefit from decades of astronomical observation of similar stars and, since about 1920, informed analysis of stellar interiors, whereas we still lack access to other Earths: extra solar planets were not observed—though much imagined—until 1992, and little is known of their present state, let alone their evolution.

Even without instructive analogy with other suns, the long view should allow us to document slow processes or detect variations in processes otherwise assumed to be constant. A measure badly in need of amplification for both purposes is solar radius. Discussion of models of the Sun's interior routinely cites luminosity and radius as two critical measures. Current technology would find it difficult to detect the rate of change estimated by helioseismology, namely 10 milliarcsec (mas) between 1996 and 1998 [5]. Satellite data for 8 years suggests that any linear trend in the radius is less than 14.5 mas/yr [13]. Unfortunately cumulative changes cannot be gauged because comparison with early measurements (and by early we merely mean the seventeenth century) are compromised by problems of inter-pretation, including confusion between visible and seismic radius [14], instru-mental inconsistencies, and—as the classical observers were all Earthbound—atmospheric distortion.

Greater success may attend the proposal to monitor solar luminosity by mon-itoring the long-term production of neutrinos in the Sun's core. Neutrinos are generated by thermonuclear reactions at an estimated rate of 10^{38}/s. Attempts to measure the flux using various detectors had repeatedly conflicted with the expectations of the Standard Solar Model (SSM) of the solar interior, and among the explanations offered were a number which focused on variations in the solar-neutrino flux that were archived in the geological record. The flux is very sensitive to the Sun's central temperature: for boron-8 (^{8}B) neutrinos, for example, the relationship is to the 25th power of the temperature; any significant variation would thus have broad implications for the SSM as well as shedding light on the 'neutrino problem.'

The obvious way forward is to identify isotopes in the crust whose evolution can plausibly be attributed to reactions with solar neutrinos. The need is therefore for a suitable parent and daughter pair of which the daughter nuclide must be sufficiently long-lived to integrate over the period of interest (long enough to yield an adequate count but, one might add, short enough to reveal variations), detection which is not compromised by stable isotopes of the same element, and a target deposit which is insulated from background or cosmic radiation, sealed, and well dated [23].

One suggestion was to put this into effect by using technetium-97 (^{97}Tc) and technetium-98 (^{98}Tc) in molybdenite (MoS$_2$) from the Red Mountain mine in Colorado at a depth of 1130–1500 m [1]. ^{97}Tc is produced by the reaction of high energy boron-8 (^{8}B) neutrinos with molybdenum-97 (^{97}Mo), ^{98}Tc by the reaction of ^{8}B with ^{98}Mo. The expectation was to find 10^8 atoms of ^{98}Tc in 50 tons of the mineral. A later proposal was to exploit the fact that neutrino interaction will convert tellurium-126 (^{126}Te) into iodine-126 (^{126}I) some of which decays into stable xenon-126 (^{126}Xe), whereupon the concentration of ^{126}Xe in rocks con-taining tellurium, such as some gold deposits, would indicate the neutrino flux since the rocks formed or were last molten. One could then compare the results for tellurides of different age [10].

Unfortunately the molybdenum project ran into a familiar problem: back-ground. Processing of ore obtained from older, shallow mining had left a 'mem-ory' in the processing plant and commercial economics ruled out adequate, processing of deep ore exclusively for the neutrino project [11]. But progress in counting of small numbers of atoms means that, with table-top 'smelting', the idea may be revisited before long.

The original enquiry had gained added motivation from a desire to account for the Earth's climatic history. The possibility that solar luminosity underwent quasi-periodic reductions by 10–30 % in luminosity on a timescale of ~ 200 million years [4] was inevitably linked to the major glaciations of the last Gyr. The notion had been narrowed down to the last glacial episode by the proponents of the ^{98}Te model experiment [1]. More generally, the editor of *Nature* [15] saw the ^{126}Te scheme as an opportunity to challenge a central assumption in the prevailing astrophysical model of the Sun: that the hydrogen-burning core is thermally stable. If incorrect, he suggested, 'there could well have been uncounted variations of solar constant' in the past. One can recognize in all these responses the long arm of cyclic interpretations. The detection of cumulative change would surely be of even greater significance in the study of the Sun's interior.

References

1. Cowan GA, Haxton WC 1982 Solar neutrino production of technetium-97 and technetium-98. Science 216:51–54. doi: 10.1126/science.216.4541.51
2. Damon PE, Laut P (2004) Pattern of strange errors plagues solar activity and terrestrial climatic data. EOS 85:370–374
3. Darwin GH (1879) On the precession of a viscous spheroid and on the remote history of the earth. Phil Trans R Soc 170:447–530
4. Dilke FWW, Gough DO (1972) The solar spoon. Nature 240:262–294. doi: 10.1038/240262a0
5. Emilio M, Kuhn JR, Bush TI, Scherrer O (2000) On the constancy of the solar diameter. Astrophys J 543:1007–1010
6. England P, Molnar P, Richter F (2007) John Perry's neglected critique of Kelvin's age for the earth: a missed opportunity in geodynamics. GSA Today 17. doi: 10.1130/GSAT01701A.1
7. Friis-Christensen E, Lassen K (1991) Length of the solar cycle: an indicator of solar activity closely associated with climate. Science 254:698–700
8. Fröhlich C, Lean J (2004) Solar radiative output and its variability: evidence and mechanisms. Astron Astrophys Rev 12:273–320. doi: 10.1007/s00159-004-0024-1
9. Gough DO (1988) Theory of solar variation. In: Cini Castagnoli G (ed) Solar-terrestrial relationships and the Earth environment in the last millennia. Proceedings of International School of Physics Enrico Fermi. North-Holland, Amsterdam
10. Haxton WC (1990) Proposed neutrino monitor of long-term solar burning. Phys Rev Lett 65:809–812
11. Haxton WC (1995) The solar neutrino problem. Annu Rev Astron Astrophys 33:459–503
12. Huxley TH (1869) Geological reform. Coll. Essays 8:305–3039
13. Kuhn JR, Bush RI, Emilio M, Scherrer PH (2004) On the constancy of the solar diameter: II. Astrophys J 613:1241–1252
14. Lefebvre S, Kosovichev AG, Rozelot JP (2007) Seismic test of nonhomologous solar radius change with the 11 year activity cycle. Astrophys J 658:L135–L138
15. Maddox J (1990) Towards a history of the Sun? Nature 346:695
16. Momigliano A (1981) The rhetoric of history and the history of rhetoric: on Hayden White's Tropes. In: Shaffer ES (ed) Comparative criticism: a yearbook. Cambridge University Press, Cambridge 3:259–268
17. Perry J (1895) On the age of the earth. Nature 51:224–227
18. Richter FM (1986) Kelvin and the age of the earth. J Geol 94:395–401

19. Solanki SK, Usoskin IG, Kromer B, Schüssler M, Beer J (2004) Unusual activity of the Sun during recent decades compared to the previous 11,000 years. Nature 43:1084–1087

20. Thomson W (1862) On the age of the sun's heat. Macmillan's Mag 5:288–293

21. Usoskin IG, Solanki SK, Schüssler M, Mursula K, Alanko K (2003) Millennium-scale sunspot number reconstruction : evidence for an unusually active Sun since the 1940s. Phys Rev Lett 91. doi: 10.1103/PhysRevLett.91.211101

22. Willson RC, Mordvinov AV (2003) Secular total solar irradiance trend during solar cycles 21–23. Geophys Res Lett 30: 1199. doi:10.1029/2002GL016038

23. Wolfsberg K, Kocharov GE (1991) Solar neutrinos and the history of the Sun. In: Sonett CP, Giampapa MS, Matthews MS (eds) The Sun in time. University Arizona Press, Tucson AZ

Chapter 2
Origins

Abstract Modern estimates of the Sun's age are based primarily on two sources: the isotopic age of components of chondritic meteorites (particularly calcium-aluminium-rich inclusions or CAIs) which are thought to have originated in the solar nebula, and the implications of the Standard Solar Model (SSM), for which helioseismology now provides corroboration. The generally accepted result is ~ 4.57 Gyr. The protoSun joined the Main Sequence, as depicted on Hertzsprung-Russell diagrams of stellar luminosity and temperature, about 40 Myr later, when its luminosity had supposedly fallen to 70 % of that characterising Main Sequence stars with a mass similar to that of our Sun. Modern counterparts of the protoSun include naked T Tauri and FU Orionis stars; ^{21}Ne from chondrites appears to indicate flare activity on it 100–1000 times the present level.

Keywords ProtoSun · Hertzsprung-Russell diagram · Main Sequence · Helioseismology · T Tauri · Chondrite · CAI · Standard Solar Model

The age of the Sun is now known from two main sources, the age of minerals in meteorites which are thought to have condensed from the dusty gas cloud that gave rise to the solar system, and numerical modelling of the processes operating in the solar interior. Both methods, first employed half a century ago, embody weighty assumptions but they yield answers which agree to a persuasive degree.

The first successful attempt to date a mineral based on radioactive decay was made by Arthur Holmes in 1911. Using the lead-uranium method, Holmes [1] obtained an age of 1640 Myr for Pre-Cambrian samples from Ceylon and thus showed that our planet must be even older. The discovery of isotopes by Frederick Soddy in 1913 [2] then led to the development of techniques which allowed a wide range of terrestrial rocks to be dated.

C. Vita-Finzi, *Solar History*, SpringerBriefs in Astronomy,
DOI: 10.1007/978-94-007-4295-6_2, © The Author(s) 2013

Fig. 2.1 Narre Gneiss Terrane, Jack Hills (WA), source of zircon dating from 4.4 Ga ago. PIA12064 courtesy of NASA/GSFC/METI/ERSDAC/JAROS and US/JAPAN ASTER Science Team. The image covers an area of $\sim 27 \times 34$ km

Meteoritic Ages

The oldest known rocks exposed at the Earth's surface are the Acasta gneiss of NW Canada, with an age of about 4.03 Gyr [3], and the Nuvvuagittuq greenstone of Quebec with a samarium-146-neodymium-142 (^{146}Sm–^{142}Nd) age of 3.8 Gyr and perhaps as much as ~ 4.28 Gyr [4], an age likely to be revised upwards by recent advances in ^{146}Sm–^{142}Nd dating [5]. Even older is a derived zircon with an age of 4.36 Gyr reported from the Jack Hills in western Australia [6] (Fig. 2.1).

But from the standpoint of solar dating the Earth plays a more important role as an accessible repository of meteorites. Initially meteorite ages served to date the Earth. A durable estimate was obtained by Clair Patterson in 1956 [7]. His ^{207}Pb–^{204}Pb age of 4.55 $\pm$ 0.07 Gyr used leads from two iron meteorites and three stony meteorites of which two were chondrites and one an achondrite. The result was subsequently revised to 4.50 $\pm$ 0.07 Gyr. Here too Patterson found it reasonable to believe that the Earth and the meteorites formed at the same time, and the age of lead from oceanic sediments supported this assumption.

Meteorites then gained prominence in attempts to date the solar system as a whole, and the analysis came to focus on the calcium-aluminium-rich inclusions

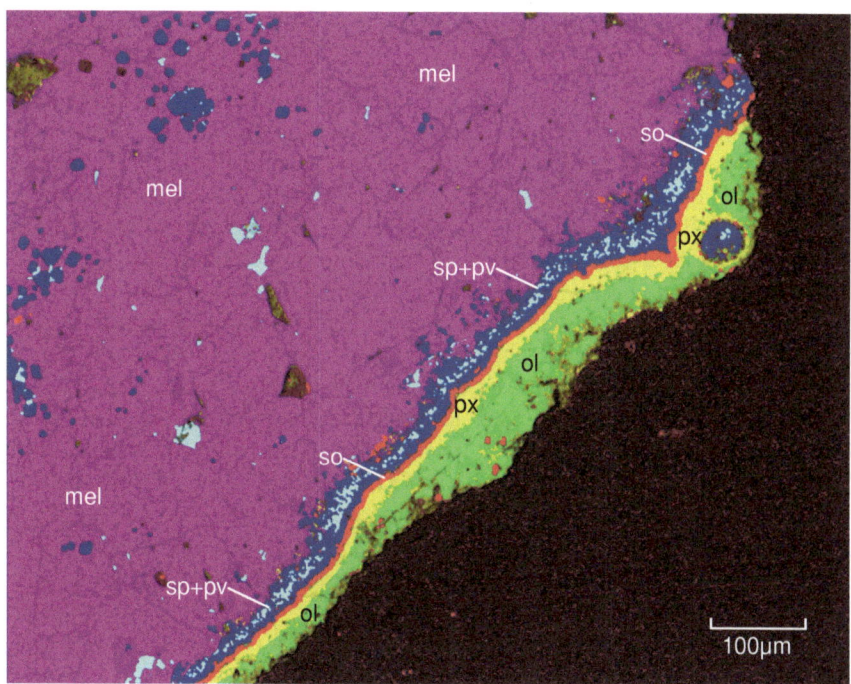

Fig. 2.2 Compositional X-ray image of rim and margin of a CAI from the Allende meteorite (courtesy of J. I. Simon and NASA). Oxygen isotope variations point to short-lived fluctuations of the environment in which CAIs formed, either because of transport of the CAIs themselves to distinct regions of the solar nebula or because of varying gas composition near the protoSun. *ol* = olivine, *mel* = melilite, *sp* + *pv* = spinel + perovskite, *so* = sodalite, *px* = pyroxene

(CAIs) found within chondritic meteorites. The isotopic composition of CAIs bears on the dynamics of the nebula in the environs of the protoSun [8] (Fig. 2.2). CAIs are thought to represent the oldest solids in the solar system and to have formed before the chondrules that give chondrites their name whereupon they were removed from the innermost protoplanetary disk. For example, the ^{207}Pb–^{206}Pb age of CAIs from the Allende meteorite is 4.565 Gyr, $\sim$ 1.66 Myr older than its chondrites [9].

The CAI maximum will doubtless continue to be revised. In 2010, for example, a CAI from the meteorite NWA 2364, which fell in Morocco in 2004, gave a ^{207}Pb–^{206}Pb age of 4.568 Gyr [10], 0.3–2 Myr earlier than existing estimates. The issue is not simply of solar system age. Thus the abundance of shortlived isotopes such as iron 60 (half-life $\sim$ 2.6 Myr) was by implication doubled by the NWA 2364 result, which in turn supported the suggestion that a supernova explosion triggered the solar system or at any rate promoted differentiation of its constituents [e.g. 11]. And the close agreement between CAI ages from several meteorites, coupled with the consistent relationship between CAI and chondrite ages, endorses the conclusion that accretion of the protoSun was part of a process already at work 4.57 Gyr ago.

Modelling the Main Sequence Sun

Arthur Eddington [12] had shown that, in accordance with Einstein's arguments, the transformation of hydrogen to helium would liberate energy. Hans Bethe fleshed out the idea as the CNO cycle and the *pp* chain, and he concluded that the former would operate in stars significantly heavier than the Sun and the latter in lighter stars [13]. The acceptance by 1932 that the Sun was rich in hydrogen thus revealed a new heat source which would not simply prolong the Sun's assumed cooling history but conceivably reverse any cooling trend at the outset.

Indeed, the hydrogen–helium model went beyond mere heat generation because it formally demoted the Sun to one of many stars, a step which brought with it the overtones of family history and thus of family destiny. That there is a multitude of other suns is of course a hoary idea advanced by among others Giordano Bruno, Descartes and William Herschel, but it was not formalised until Eynar Hertzsprung (in 1911) and Henry Norris Russell (in 1913) plotted the relationship between spectral type and luminosity (Fig. 2.3). Interpretation of the diagram, and indeed the allocation of stars to their individual places on it, hinge on a number of their own assumptions, notably the links between spectral type and temperature or between luminosity and some measure of magnitude such as apparent magnitude at 10 pc (parsec or $\sim$3.3 light-years), and some H-R diagrams simply employ absolute visual magnitude and the B–V colour index, which is the difference in magnitude determined at two standard wavelengths (B = 440, V = $\sim$550 nm). The crucial finding was that about 85 % of observed stars lie in a diagonal band known as the Main Sequence (MS) and the remainder fall into groupings now known as white dwarfs, red dwarfs, giants and super giants. The Sun features halfway along the MS for nearby stars.

The H-R diagram put the Sun in its place as a run-of-the-mill star. But it did more than demystify the Sun. Combined with observational and theoretical data and calculations it yielded a powerful genealogical message. Stars evolve and their evolution can be traced by the ergodic principle that sampling in space may be equivalent to sampling in time. Solar history could now be written by a combination of calculation and analogy: calculation because the present state of the Sun—insofar as its internal temperature could be surmised—was the outcome of a physical process operating at a knowable rate acting on a stock of fuel proportional to the mass of the star, and analogy because the Sun at different stages in its life could be matched with other stars on the HR diagram.

This is not the same as assuming that a star will progress up or down the MS slope. The MS merely represents what Lewis [14] called the stable locus occupied by stars of different masses. Thus plotting mass against luminosity for nearby stars for which the two attributes are well documented gives a relationship of the form L $\alpha\ M^{3.5}$ for stars with a mass greater than a few tenths of $M_\odot$, and L $\alpha\ M^2$ for smaller stars.

The pre-MS Sun is often compared with T Tauri and FU Orionis stars. T Tauri (Fig. 2.4), which gave its name to a class of irregular variable stars, was

Fig. 2.3 Schematic
Hertzsprung-Russell diagram
showing major stellar
categories derived from
correlation of luminosity with
surface temperature.
Conventional nomenclature
for spectral classes is shown
at the top. Note T Tauri zone
and approximate location of
stars with 2 (2M) and 10
(10M) solar masses. Various
sources

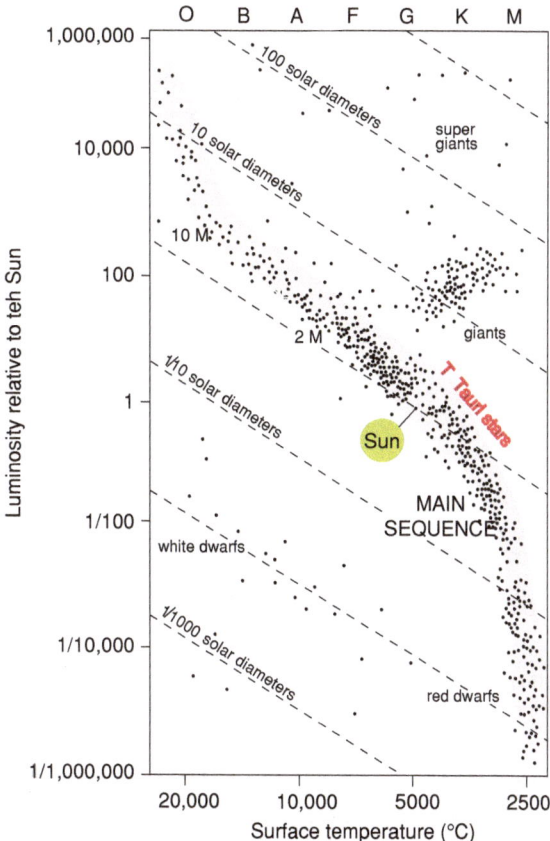

discovered in 1852. Between 1864 and 1916 it varied in magnitude irregularly
between 9.3 and 14. It now oscillates between 9.3 and 10.7. According to the
American Association of Variable Star Observers [15] it also displays variations of
a few tenths almost daily, perhaps through violent activity in its atmosphere
(Fig. 2.5) or the infall of material from its accretion disk. The crucial characteristics of classical T Tauri stars (CTTS) are recent emergence from a dust and gas
cloud, growth by accretion, masses of ~ 0.2–3 $M_\odot$, and an age of 10^5–10^6 yr [16].
Their central temperature is considered to be too low for hydrogen fusion (hence
the presence of lithium as well as hydrogen and helium in the parental gas: lithium
is destroyed when temperatures exceed 2.5×10^6 K) and their energy is gravitational energy—an interesting revival of the Kelvin–Helmholtz mechanism. The
Röntgen Satellite (ROSAT) X-ray mission (1990–1999) in combination with
ground-based optical observation revealed several hundred new T Tauri stars on
the basis of H alpha emission and lithium absorption.

Weak-line (or naked) T Tauri stars (NTTS) lack strong emission lines and an
accretion disk, and they may epitomise the Sun, which started off as a CTTS, after
its inner protoplanetary disk had collapsed [17]. Indeed, NTTS display dark spots

Fig. 2.4 T Tauri, prototype of the class of T Tauri variable stars, and a nearby yellow cloud—Hind's Variable Nebula—which may contain another young stellar object (courtesy of Adam Block, Mount Lemmon SkyCenter, University of Arizona)

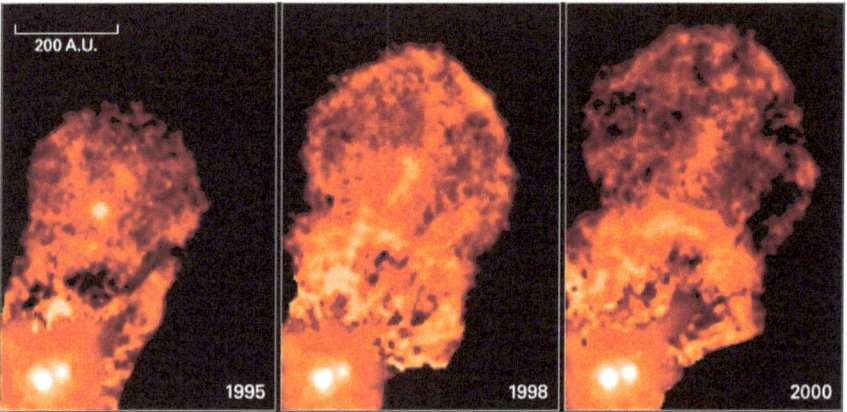

Fig. 2.5 Bubbles of gas leaving the young XZ Tauri binary star system in the Taurus star-forming region (courtesy of J. Krist (STScI), the WFPC2 Science Team and NASA)

covering 5–10 % of their surface compared with $\sim$ 1 % on the Sun, and rotate with a period of 1–10 days.

The third group of potential analogs to the early Sun are the FU Orionis stars. In 1937 the star FU Orionis was found to rise in magnitude from 16.5 to 9.6 (5 steps in apparent magnitude represent a hundredfold change in brightness; mid-visual wavelength of ~ 555 nm is assumed). FU Orionis now has a magnitude of 8.9. Other stars as indecisive about their magnitude and spectral type were later discovered, including V1057 Cygni, and FU Orionis stars may represent a stage in the development of T Tauri stars rather than a violent subclass. In 2011 over a period of 13 months, a classical T Tauri star acquired the characteristics of an FU Orionis object (PTF 10qpf), perhaps during a period of enhanced disk accretion [18].

If we focus on postulated pre-main sequence (MS) stars of mass similar to our Sun's we can infer an initial phase of contraction to produce a large, luminous (radius $> 50\ R_\odot$, luminosity 150 $L_\odot$) protoSun in hydrostatic equilibrium. Once the effective temperature has risen sufficiently the process of hydrogen fusion begins. The Sun joined the zero-age MS after about 40 Myr; its luminosity was 0.7 $L_\odot$ [19]. If we accept that the accretion disk, meteorite formation and the emergence of the central star occurred within 10 Myr of each other, the time to zero age on the MS (ZAMS) is 40 $\pm$ 10 Myr and the Sun's age is ~ 4.53 Gyr [20].

There are two important corollaries. The active T Tauri Sun presumably emitted a powerful solar wind more powerful than the present, which currently attains energies of 1.5–10 keV and travels at average speeds of 400 km/s. The Sun would also have been highly luminous in the UV wavelengths [14].

Both features offer the promise of geological detection. Evidence that the Sun passed through a T Tauri phase before joining the MS has been obtained from the rare gas content (notably neon-21 (^{21}Ne)) of grains from a number of a class of gas-rich meteorites, among them the carbonaceous chondrites Murchison, Murray and Cold Bokkeveld, the chondrite Fayetteville, and the gas-rich achondrite Kapoeta. If we except analogy with other stars, ^{21}Ne in meteorites at present provides 'practically the only way we have to study the possible existence of a T Tauri stage of the Sun' [21]. The isotopes were produced by spallation (that is, the expulsion of protons and neutrons) caused by irradiation by energetic protons. The process must have operated before the meteorite was compacted, as heavy ions from solar flares and those implanted by the solar wind penetrate respectively <1 mm and <1 μm (see Chap. 3). The grains were presumably then shielded from further irradiation by burial until the parent body was exposed to further solar irradiation and galactic cosmic rays (GCRs) en route to Earth. GCRs consist of about 89 % protons or hydrogen nuclei, 10 % alpha particles or helium nuclei and 1 % nuclei of heavier elements.

The early proton flux appears to have exceeded that of modern solar flares by several orders of magnitude. Moreover the neon isotopes indicate a harder energy spectrum, that is to say one with higher energy levels, higher frequency and shorter wavelength, but solar flare activity much higher than on the Sun today [22] is postulated on the unproven grounds that exposure was shortlived [23].

Moreover, silicate grains in some stone and iron meteorites display a bimodal distribution of track densities. One fraction was presumably produced (like the rare gases) at a time when the grains were unshielded, and it indicates irradiation by an intense Sun. The other fraction arose when the meteorites, by now compacted, were exposed to GCRs in interplanetary space (see this chapter).

Fig. 2.6 Internal rotation rate
(*red*: fast, *blue*: slow) derived
from helioseismology.
Dashed line marks the
tachocline at the base of the
convection zone (courtesy of
M J Thompson and NASA)

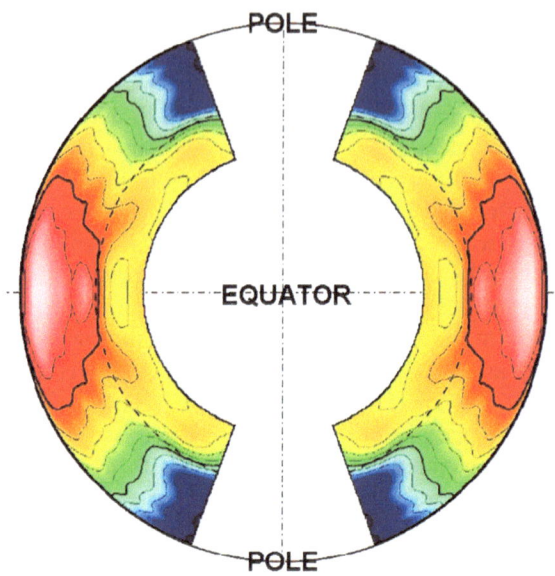

The Contribution of Helioseismology

Helioseismology is the study of wave oscillations, especially acoustic pressure waves, in the Sun. The waves are transmitted to the photosphere, the Sun's visible surface, where they can be observed. Pulsation occurs in about 10 million resonant modes; each mode samples a different depth in the solar interior. The oscillations can be detected as Doppler shifts of lines in the solar spectrum. They were recorded from 1995 on the Solar and Heliospheric Observatory (SOHO) satellite as part of the solar oscillations investigation using the Michelson–Doppler Imager (SOI–MDI) before this was superseded in 2011 by the HMI–SDO, the Heliose-ismic and Magnetic Imager on the Solar Dynamics Observatory. Observations from space are complemented by data from the six ground stations of the Global Oscillation Network Group (GONG), which ensures essentially continuous observation of solar oscillations.

Analysis of several thousand wave modes in the 5 min period range yield information about static and dynamic properties of the Sun's core and its convection zone (Fig. 2.6). The two main kinds of information that result are spatial averages of the speed of travel of 'seismic' waves and spatial averages of relative motion within the Sun, but increasingly local features such as flow beneath sunspots and cell convection are being analysed. In the present context the crucial finding is the proportion of helium to hydrogen and thus the time during which the fusion process has operated.

Previously helium content, which could not be determined satisfactorily using spectroscopy, was estimated from solar luminosity. For instance, a model of the solar interior based on the p–p chain which included a convective zone and the effects of the conversion of hydrogen to helium and which assumed a heavy element abundance of 0.04, a central temperature of 15.8×10^6 K and a central pressure of 127 g cm^{-3}, yielded the observed solar luminosity and radius if it postulated a helium abundance of 0.26 [24]. Helioseismic inversion of solar oscillation (p-mode) frequencies then gave a helium abundance of ~ 0.25 or ~ 0.23 in the convection zone according to which equation of state was adopted. In both cases the helium abundance in the solar corona and the solar wind is substantially lower, which suggests that element separation occurs in the solar atmosphere as well as in the solar interior [25].

The uncertainties over the Sun's helioseismic age will inevitably shrink but it cannot be excluded that data from other sources, including the Earth and the Moon, will call for some of the critical values to be reconsidered. They include such variables as metal abundance and opacities (that is, a measure of the resistance presented to the free passage of photons) in the interior of the Sun and complications such as differential settling of its constituents over time. It is instructive that, where one determination gave a helium value of 0.248 ± 0.002 and a resulting age of 4.66 ± 0.11 Gyr [26], a relativistic correction for the assumed equation of state reduced the age estimate to 4.57 ± 0.11 Gyr [27]. Similarly, the rather different ages for the onset of hydrogen burning of 4.52 and 4.6 Gyr both appeared consistent with helioseismic estimates of sound speed [28].

References

1. Holmes A (1911) The association of lead with uranium in rock-minerals, and its application to the measurement of geological time. Phil Trans Roy Soc A 85: 248–256
2. Lewis C (2011) Holmes's first date. Geoscientist 21: 12–18
3. Bowring SA, Williams IS (1999) Priscoan (4.00-4.03) orthogneisses from northwestern Canada. Contrib Mineral Petrol 134: 3–16
4. O'Neil J, Francis D, Carlson RW (2011) Implications of the Nuvvuagittuq greenstone belt for the formation of Earth's early crust. J Petrol 52: 985–1009, doi:10.1093/petrology/egr014
5. Kinoshita N and 18 others (2012) A shorter ^{146}Sm half-life measured and implications for ^{146}Sm–^{142}Nd chronology in the solar system. Science 335: 1614–1617, doi:101126/science1215510
6. Wilde SA, Valley JW, Peck WH, Graham CM (2001) Evidence from detrital zircons for the existence of continental crust and oceans on the Earth 4.4 Gyr ago. Nature 409: 175–178
7. Patterson C (1956) Age of meteorites and the earth. Geochim Cosmochim Acta 10: 230–237
8. Simon JI, Hutcheon ID, Simon SB, Matzel JEP, Ramon EC, Weber PK, Grossman L, DePaolo DJ (2011) Oxygen isotope variations at the margin of a CAI records circulation within the solar nebula. Science 331: 1175–1178. doi:10.1126/science.1197970
9. Connelly JN, Amelin Y, Krot AN, Bizzarro M (2008) Chronology of the solar system's oldest solids. Astrophys J 675:L121–L124
10. Bouvier A, Wadhwa M (2010) The age of the Solar System redefined by the oldest Pb–Pb age of a meteoritic inclusion. Nature Geosci 3:637–641
11. Cameron AGW, Truran JW (1977) The supernova trigger for formation of the solar system. Icarus 30:447–461
12. Eddington A (1920) Space, time and gravitation: an outline of the general relativity theory. Cambridge University of Press, Cambridge

13. Bahcall JN, Salpeter EE (2005) Stellar energy generation and solar neutrinos. Phys Today 58:44–47
14. Lewis JS (1997) Physics and chemistry of the solar system. Academic, San Diego
15. American Association of Variable Star Observers (2012) at http://www.aaso.org/vsots_ttau
16. Cohen M (1981) Are we beginning to understand T Tauri stars? Sky Telescope 61:300–303
17. Walter FM, Barry DC (1991) Pre- and main-sequence evolution of solar activity. In: Sonett CP, Giampapa MS, Matthews MS (ed) The Sun in time. University of Arizona, Tucson
18. Miller AA and 30 others (2011) Evidence for an FU Orionis-like outburst from a classical T Tauri star. Astr J 730: 80. doi:10.1088/0004-637X/730/2/808
19. Bhandari N (1997a) Paleoactivity of the Sun: meteoritic perspective. In Cini Castagnoli G, Provenzale A (ed) Past and present variability of the solar-terrestrial system. IOS, Amsterdam
20. Demarque P, Guenther DB (1999) Helioseismology: probing the interior of a star. Proc Nat Acad Sci 96:5356–5359
21. Lal D, Lingenfelter RE (1991) History of the Sun during the past 4.5 Gyr as revealed by studies of energetic solar particles recorded in terrestrial and extraterrestrial samples. In: Sonett CP, Giampapa MS, Matthews MS (eds) The Sun in time. University of Arizona, Tucson
22. Caffee MW, Hohenberg CM, Swindle TD, Goswami JN (1987) Evidence in meteorites for an active early Sun. Astrophys J 313:L31–L35
23. Wood PA, Pellas P (1991) What heated the parent meteorite planets? In: Sonett CP, Giampapa MS, Matthews MS (eds) The Sun in time. University of Arizona, Tucson
24. Ulrich RK, Cox AN (1991) The computation of standard solar models. In: Cox AN, Livingston WC, Matthews MS (eds) Solar interior and atmosphere. University of Arizona, Tucson
25. Kosovichev AG (1995) Inferences of element abundances from helioseismic data. In: Ulrich RK et al (ed) GONG'94: helio- and asteroseismology from the earth and space. ASP, San Francisco
26. Dziembowski RO, Sinkiewicz R, Goode PR (1998) In: Korzennik SG, Wilson A (eds) The structure and dynamics of the interior of the Sun and Sun-like stars. European Space Agency, Noordwijk
27. Bonanno A, Schlattl H, Paternò L (2002) The age of the Sun and the relativistic corrections in the EOS. Astron Astrophys 390:1115–1118
28. Brun AS, Turck-Chièze S, Morel P (1998) Standard Solar Models in the light of new helioseismic constraints. I. The solar core. Astrophys J 506:913–925

Chapter 3
The Young Sun

Abstract Meteorites and the Moon yield information on solar flares and the solar wind which, though patchy, spans 10^9 yr. However, as the solar wind and flares are primarily creatures of the corona they are poor measures of solar activity as a whole. According to the Standard Solar Model the Sun's luminosity has increased from 0.7 to 1.0 $L_\odot$ in the course of the last 4.5 Gyr. The widespread presence of liquid water on the Earth's surface at least during the last 3.8 Gyr conflicts with this finding. Various mechanisms have accordingly been proposed for warming the early Earth, including atmospheric greenhouse effects, variations in cloud cover, and the prevalence of oceans with a low albedo. Changes in the mass of the early Sun and in its activity have also been suggested in order to reconcile the SSM with the geological evidence on Earth.

Keywords Solar flares · Moon cores · Faint young sun · Greenhouse gases · Albedo

The history of the Sun once it had joined the Main Sequence is generally derived from the SSM, according to which hydrogen in the Sun's interior is progressively transformed into helium. This process leads to increased density and temperature, and consequently also luminosity.

Solar luminosity refers to radiant energy emitted in all directions; total solar irradiance (TSI), formerly termed the solar constant, is the radiation per unit area incident on the Earth's upper atmosphere at a distance of one astronomical unit (AU) from the Sun and includes electromagnetic radiation (EMR) at all wavelengths. Many have argued that TSI variations are a consequence of surface magnetic features such as sunspots and faculae, with blocking of radiated energy by sunspots a key element and some contribution by major flares, but it is likely that the Sun's internal magnetic field dominates long-term variations in TSI [1].

However, different wavelengths originate preferentially, though not uniquely, in different parts of the Sun (see Table 3.1), and it would therefore pay to

Table 3.1 Solar EMR

Wave band	Wavelength	Main originating zone
Infrared	700–1,000 nm	Chromosphere
Visible	400–700 nm	Photosphere
UV	100–400 nm	Chromosphere-corona transition
X-rays	0.1–100 nm	Corona, flare
Gamma rays	10^{-6}–0.1 nm	Flare

disaggregate the processes in order to add detail to an account of the Sun's evolution on the MS. Moreover the processes in question greatly differ dynamically: photons from the solar radiative zone may require 10^5–10^6 years to reach the photosphere whereas X-ray and gamma ray formation by solar flares may take a few minutes at most to materialize. Images of the Sun at different wavelengths thus do not reflect synchronous activity in all parts of the Sun, and even if solar history has been manifested as a monotonic increase in luminosity there were doubtless changes in tempo, as noted in Chap. 1.

The Moon and Meteorites

Over the aeons the unconsolidated deposits making up the regolith on the Moon have been affected by radiation from the Sun and from space. The outcome is track production by solar-flare heavy nuclei, damage to the crystalline structure by the solar wind, isotopic changes produced by interaction of solar protons and α particles, and chemical and isotopic change caused by implantation of solar wind and solar flare ions [2].

The first and probably still the strongest flare that has been documented was observed almost by accident by Richard Carrington in 1859 when he was engaged in the daily task of plotting sunspots. His observation (necessarily in white light), though cut short by the search for someone else to witness it, sufficed to show that the phenomenon was in some way associated with sunspots. It soon became clear from interference with telegraphy that it also impinged on the Earth's magnetic environment.

All flares are explosive and the energy they release varies over at least 8 orders of magnitude (up to 10^{25} J), their height ranges between $<10^4$ and 10^5 km, and their duration between 10^3 and 10^4 s; they have been observed at wavelengths ranging from radio to gamma rays [3]. The release by a single flare of 10^{25} J of energy appears trivial only when compared with the Sun's total energy output of 10^{26} J/s. But in the present context interest resides in any evidence for a long-term increase or decrease in the frequency and size of flare events in their own right and also as surrogates for the $\sim$11-year (Schwabe) solar cycle: major flares occur perhaps once a week and weak flares a few times a day during solar maximum [4].

The energy of flares before its release is stored in the Sun's corona, so that their relation to the solar interior is very indirect. But flares usefully supplement the limited observational data of coronal activity obtained during real or simulated eclipses. Deposits on the Moon were inevitably affected by solar flares and other bursts of radiation from the Sun once the magma ocean that resulted from the originating impact had solidified. Particles derived from solar flares are sometimes included in the category of solar cosmic rays (SCR) together with solar energetic particles (SEPs) which have been accelerated to energies of up to the GeV level. They too consist of 98 % protons and penetrate into lunar material by up to 1 cm compared with 1 μm by solar wind particles. So far as we know the Moon has never had an appreciable atmosphere, and therefore has experienced neither contamination by gases nor erosion and deposition by fluids, but hints of an early global magnetic field [5] show that the evidence may have been distorted by magnetic shielding.

Moreover the record is unequal between the near and far lunar hemispheres, as the former is shielded by the Earth's magnetic field during part of its orbit around the Sun, an effect which was only acquired once the lunar rotation was tidally synchronised but now amounting to a threefold difference in response to the solar wind. Continuous measurements by the Solar Wind Spectrometer on the Moon showed that, when outside Earth's magnetic field, proton density was 10–20 per cm^3, with most protons having velocities between 400 and 650 km/s. During the five days of each month when the Moon was inside the Earth's geomagnetic tail, and during the lunar night, no solar wind particles were detectable.

Distorted or not, the lunar flare evidence is far longer than the Earth's, that is to say 10^9 years as opposed to the 10^2 years of dependable direct observation. Admittedly it is difficult to decipher, because most of the samples had more than one exposure to the radiation as a consequence of reworking ('gardening') of the loose surface deposits or regolith by meteorite impacts and occasional landslides, and have also enjoyed periods sheltered beyond the ~ 1 mm range of solar energetic particles [6]. Drilling by astronauts up to 3 m into consolidated fragments (breccia) (Fig. 3.1) has yielded samples that were exposed up to a Gyr ago at rates that can only be averaged over periods of 10^4–10^6 year [6]. A case has been made for a return to the Moon by astronauts precisely because they could locate and sample ancient, buried regoliths [7] in order to subdivide the lunar evidence into successive dated intervals.

The flare evidence yielded by meteorites has been acquired on the journey to Earth and, thanks to geomagnetic shielding, to a lesser degree once the meteorite has landed. We have already seen (in Chap. 2) that, despite these limitations, pre-compaction grains in some gas-rich meteorites yield heavy ion tracks (Fig. 3.2) and spallogenic neon-21 (^{21}Ne) which is taken by some investigators to represent enhanced solar-flare activity 2–4 orders of magnitude greater than for the present-day Sun and thus support the T Tauri model of the nascent Sun.

Fig. 3.1 Astronaut Alan L. Bean, lunar module pilot, drives a core sample tube into the lunar surface during the Apollo 12 extravehicular activity. Date 20.11.1969, Apollo XI. Photo AS12-49-7286 (courtesy of NASA). An electric drill was used on Apollos 15, 16 and 17 and it collected cores up to 3 m long

Fig. 3.2 Track density profiles in lunar rocks and meteorite grains exposed to solar flares. The similarity in the slope of the track density profiles for a Surveyor glass filter exposed on the Moon during 1967–1970 (SG, $\times 10^5$), lunar soil (LS; b $\times$ 10), rocks (LR, c $\times$ 100, last 10^4–10^6 yr) and breccias (LB, >3.9 Gyr ?), and grains from the Fayetteville F, >4 Gyr) and Murchison (M $\times$ 10, > 4.4 Gyr) meteorites is taken to signify that the energy spectrum for heavy ions from large solar flares has remained broadly similar over the last few Gyr (adapted from [6])

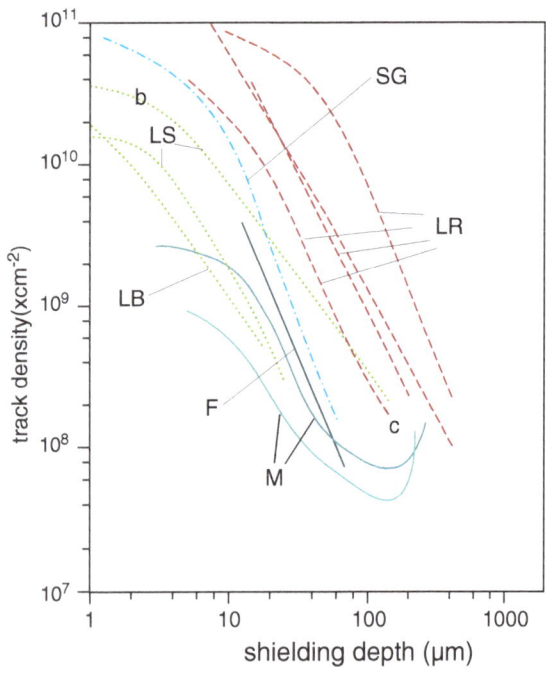

The Faint Young Sun

One of 'the most unavoidable consequences of stellar structure considerations' [8] is that the luminosity of the Sun when it joined the MS was about 70 % (0.7 $L_\odot$) of today's value, when it is classified as a G (spectral class) V (luminosity) star. The estimate embodies a number of assumptions, notably hydrostatic balance throughout the Sun's MS life, spherical symmetry, no rotation, no magnetic fields, initial chemical homogeneity, thermal balance, constant mass, and mixing confined to the convection zones [9].

One might expect events on Earth to chime with the progressive rise in luminosity. As Martin Schwarzschild commented in 1958 [10 n.7] 'Can this change in the brightness of the sun have had some geophysical or geological consequences that might be detectable?'

According to Sagan and Mullen [10], with an initial solar luminosity 30 % below the present, under current terrestrial albedo and atmospheric composition, and with an infrared emissivity of 0.9, global mean temperatures 4–4.5 Gyr ago were about 263 K and did not exceed the freezing point of sea water until 2.3 Gyr ago. (Today's global mean is 286–288 K.) But the evidence of geology and palaeontology pointed to abundant liquid water 4.0–2.7 Gyr ago, with the earliest microfossils dated from 3.2 Gyr and algal stromatolites in many parts of the world from 2–2.8 Gyr.

These figures have changed slightly since 1972: 3.4 Gyr is now favoured for the oldest microfossils [11], and there is evidence of a hydrosphere perhaps as early as 4.4 Gyr ago [12]. But if anything the new values heighten the discrepancy. And the picture has been confused by evidence that the Earth experienced a phase of intense meteoritic bombardment about 3.9 Gyr ago [13], and, more controversially, that successive global glaciations ('snowball Earth'), the oldest 2.4–2.1 Gyr ago and the latest ~650 Myr ago, were promoted by a runaway ice-albedo effect until cut short by high levels of atmospheric CO_2 [14].

Setting these complications aside, the discrepancy between the luminosity model and the field evidence had to result from an error in the initial luminosity, the albedo, or atmospheric composition [10]. The first had been put at the conservative value of 30 % and any increase could only make the conflict worse. Satellite data had indicated an albedo of 0.3; Sagan and Mullen [10] adopted 0.35 to secure agreement with their observed surface temperature, but a reduction in albedo to raise the modelled temperature seemed to be ruled out especially as low temperatures were likely to increase the area covered by ice and snow with albedos of 0.5–0.7. That left as the only option a change in atmospheric composition (Fig. 3.3).

The need was for a gas with significant absorption in the middle infrared near the Wien peak (i.e. the wavelength such that half the total radiation intensity falls on either side of it) of Earth's thermal emission, namely 8–13 μm, and Sagan and Mullen [10] lit upon ammonia (NH_3), a gas which is thermodynamically unstable in the present-day atmosphere but which might have survived in the low O_2 levels that prevailed for much of early Earth history. To be sure, ammonia is easily broken down by UV in oxygen-free conditions, but in late Archaean times (3.8–2.5 Gyr),

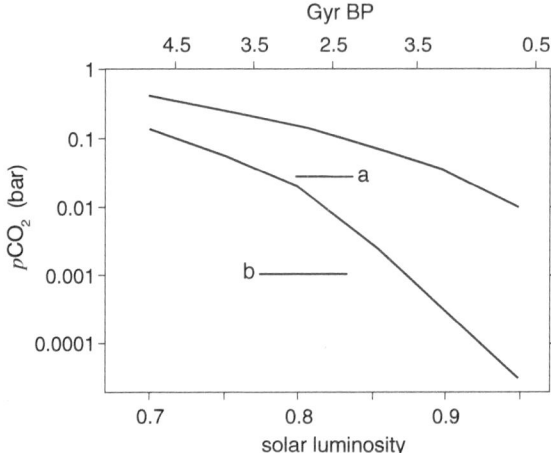

Fig. 3.3 The curves show estimated pCO_2 values required to keep Earth's average surface temperature at freezing point (273 K) or at today's value (288 K) with present albedo and CO_2 and H_2O the only important greenhouse gases. Estimated lower limit of pCO_2 based on siderite in banded iron formations *a* after [35] and estimated upper limit also based on siderite *b* after [24]; *c* = today's pCO_2. Relationship between age (top scale) and solar luminosity relative to today after [9]. Simplified after [36]

at least, it might have been shielded by a photochemical haze composed of hydrocarbon aerosols [15].

In the event, ammonia found little favour and was supplanted by methane (CH_4), which could have been contributed by methanogenic organisms [16]. In combination with nitrous oxide (N_2O) it could have produced up to 10° of warming in the Proterozoic 2.5–0.54 Gyr ago [17] except when increased O_2 concentration reduced production of the two gases by organisms and triggered glaciations. Likewise, a cocktail of CH_4-CO_2-H_2O could have been helped by ethane (C_2H_6) to keep the late Archaean warm until rising O_2 levels introduced glacial conditions [18].

Nevertheless CO_2, the obvious contender especially in view of what we know about the runaway greenhouse on Venus, has its proponents, some of whom argue that the dependence of silicate weathering on temperature results in a negative feedback between atmospheric CO_2 and surface temperature: a fall in temperature leads to reduced weathering and to CO_2 accumulation in the atmosphere [19]. But the CO_2 model has been rejected among other things on the grounds that, as the strongest bands are nearly saturated, even high levels would have little effect. In any case the siderite ($FeCO_3$) that would probably accumulate in an atmosphere rich in CO_2 is not common in the rocks laid down 2.8–2.2 Gyr ago [20]. Water vapour, another obvious candidate, faces the problem that a greater level in the atmosphere than at present would be unlikely at a time of lower temperatures, although the combination of CO_2 + H_2O in a weakly reducing atmosphere has in fact been proposed [21].

An alternative approach has been to argue that the discrepancy between low solar luminosity and a non-glacial Earth has been exaggerated or at any rate that it can be eliminated without major atmospheric surgery. Thus a revised atmospheric model [22] suggests that, during the Archaean and Proterozoic (i.e. 3.8–0.54 Gyr ago), a partial pressure of 2.9 mb of CO_2 would have sufficed to bring average surface temperatures to 273 K (the freezing point of water), an order of magnitude less than previously proposed. Or, if the oceans were indeed ice-covered, as on Jupiter's moon Europa, they would have been repeatedly punctured by volcanoes and meteorite impacts which, in addition, might generate a variety of gases leading to the repeated development of transient methane-rich atmospheres [23]. In fact, if liquid oceans were widespread the global albedo would have been lower than with extensive land masses even in the absence of a reflective ice and snow cover [24] and more solar radiation would have been absorbed.

Even more bold is the suggestion that the young Sun had a larger mass than today: an extra 3–7 % would suffice, though more would evaporate the oceans. As the solar flux at a planet is sensitive to mass by the relation $F \propto M^{6.75}$ [25], there would no longer be a discrepancy to resolve. The loss of mass in the shape of solar-wind flux over a period of 1 Gyr might subsequently have brought the Sun's luminosity down to its present level.

The lunar record suggests that significant solar mass loss operated over no more than 0.1 Gyr, and the proposed rate in any case is far greater than that observed for solar-type stars [26]; a pity, as a vigorous solar wind would have deflected GCRs more effectively than it does today and raised global temperature by reducing the ionization that promotes low altitude cloud formation [27]. The role of the solar wind as Earth's shield against GCRs is a prominent item in the discussion of cosmogenic isotopes (Chap. 4) as well as in the analysis of recent climatic history (Fig. 3.4). Here it gains zest from the suspicion that, besides UV and X radiation, the young Sun emitted coronal mass ejections (CMEs) more frequently than nowadays. CMEs are bursts of plasma amounting to 10^{16} g or more which are accelerated to speed of up 2000 km/s. Kappa Ceti, a yellow dwarf star with a mass of 1.04 $M_\odot$ which is thought to be about 700 Myr old on the basis of its rotation period of 8.6—in other words like our Sun as it would have been 4 Gyr ago— emits abundant CMEs [28]. An increased incidence of CMEs, like a vigorous solar wind, or a shortage of biologically-induced cloud condensing nuclei in Archaean times [24] might reasonably be linked to a reduced GCR flux to Earth and thus a lower cloud-generated albedo. Its output in the UV wavelengths is put at 35 % lower than the present Sun between 210 and 300 nm but at least 2–7 x higher at 110–140 nm [29], which would have been even more damaging to atmospheric ammonia than is usually assumed.

Fig. 3.4 The Solar Wind Composition Experiment, performed on Apollo missions 11, 12, 14, 15, and 16, consisted of an aluminium foil sheet which was deployed on a pole facing the Sun for periods ranging from 77 minutes on Apollo 11 to 45 hours on Apollo 16. Variations in the composition of the solar wind were correlated with the intensity of the solar wind as determined from magnetic field measurements (courtesy of NASA (AS11-40-5873))

Table 3.2 Greenhouse gases on the early Earth[a]

Principal greenhouse gas	Source
CO_2	[18][21]
NH_3	[10]
$CO_2 + H_2O$	[20]
$CH_4 + CO_2 + C_2H_6$	[17]
CH_4	[15]
$NH_4 + N_2O$	[16]

[a] some atmospheric H_2O is invariably assumed but in one case it is an explicit ingredient

The SSM

The most radical of all standpoints is of course one which challenges the luminosity changes implicit in the SSM. Various non-standard solar models have been proposed. Some are 'strongly disfavored' by the helioseismological data but the luminosity evolution predicted by three deviant models and the SSM differ by 1 % over the period since ZAMS and that mainly in the early stages [30] and they agree with an increase in solar luminosity of 48 % rather than the 25–30 % cited by other authors. However, all the 12 models in question [30], (Table 3.2) assume very similar initial helium and heavy element mass fractions.

Recent spectroscopic estimates have prompted revision of the heavy element value (from the widely accepted ~ 0.229 to ~ 0.122) resulting in conflict with various aspects of the SSM and in particular the depth of the convective zone [31]. Some severe critics of the SSM (e.g. [32]) favour wholesale revision of internal solar composition and a role for a supernova that goes beyond that of solar system trigger to source. Their evidence is mainly the isotopic composition of meteorites; whatever its validity, the terrestrial evidence bearing on the Sun's luminosity should perhaps be treated as a primary source of information and not simply reconciled with the implications of the accepted solar model. Identifying the net receipt of different wavebands then becomes an essential component of the data.

In the meantime the relatively crude measures of geosciences have a useful role by narrowing the range, duration and variability of the luminosity on Earth or on natural and artificial satellites. The candidate greenhouse gases can be monitored not only in terms of their greenhouse effect but by other geochemical tests: as we saw, atmospheric CO_2 levels can to some extent be identified from siderite in fossil soils; they can also be traced in ice cores over the last 2 Myr [20]. And analysis of the solar cycle and its multiples may prove of interest in view of the many reports that the neutrino flux is anti-correlated with sunspots (e.g. [33]) which, though perhaps statistically flawed [34], encourage rigorous time-series analysis of neutrino flux.

References

1. Sofia S (2004) Variations of total solar irradiance produced by structural changes of the solar interior. EOS Trans Am Geophys Un 85:217
2. Bhandari N (1997) The Sun through time: the lunar and planetary perspective. In: Cini Castagnoli G, Provenzale A (eds) Past and present variability of the solar-terrestrial system. IOS, Amsterdam
3. Shibata K, Magara T (2011) Solar flares: magnetohydrodynamic processes. Living Rev Solar Phys 8:1–99
4. De Pater I, Lissauer JJ (2001) Planetary sciences. Cambridge University Press, Cambridge
5. Runcorn K (1983) Lunar magnetism, polar displacements and primeval satellites in the Earth–Moon system. Nature 304:589–596
6. Goswami JN (1991) Solar flare heavy-ion tracks in extraterrestrial objects. In: Sonett CP, Giampapa MS, Matthews MS (eds) The Sun in time. University of Arizona, Tucson
7. Crawford IA, Fagents SA, Joy KH, Rumpf ME (2010) Lunar palaeoregolith deposits as recorders of the galactic environment of the solar system and implications for astrobiology. Earth Moon Plan. doi 10.1007/s11038-010-9358-z
8. Newman MJ, Rood RT (1977) Implications of solar evolution for Earth's early atmosphere. Science 198:1035–1037
9. Gough DO (1981) Solar interior structure and luminosity variations. Solar Phys 74:21–34
10. Sagan C, Mullen G (1972) Earth and Mars: evolution of atmospheres and surface temperatures. Science 177:52–56. doi:10.1126/science.177.4043.52
11. Wacey D, Kilburn MR, Saunders M, Cliff J, Brasier MD (2011) Microfossils of sulphur-metabolizing cells in 3.4-billion-year-old rocks of Western Australia. Nature Geosci 4: 698–702
12. Wilde SA, Valley JW, Peck WH, Graham CM (2001) Evidence from detrital zircons for the existence of continental crust and oceans on the Earth 4.4 Gyr ago. Nature 409:175–178

13. Abramov O, Mojzsis S (2009) Microbial habitability of the Hadean Earth during the late heavy bombardment. Nature 459:419–422. doi:10.1038/nature08015
14. Hoffman PF, Jaufman AJ, Halverson GP, Schrag DP (1998) A neoproterozoic snowball Earth. Science 281:1342–1356. doi: 10.1126/science.281.5381.1342
15. Sagan C, Chyba C (1997) The Early faint Sun paradox: organic shielding of ultraviolet-labile greenhouse gases. Science 276:1217–1221. doi: 10.1126/science.276.5316.1217
16. Pavlov AA, Kasting JF, Brown LL, Rages KA, Freedman R (2000) Greenhouse warming by CH_4 in the atmosphere of early Earth. J Geophys Res 105:11981–11990
17. Robertson AL, Roadt J, Halevy I, Kasting JF (2011) Greenhouse warming by nitrous oxide and methane in the Proterozoic Eon. Geobiology 9:313–320
18. Haqq-Misra JD, Domagal-Goldman SD, Kasting PJ, Kasting FJ (2008) A revised, hazy methane greenhouse for the Archean Earth. Astrobiol 8:1127–1137. doi: 10.1089/ast.2007.0197
19. Kasting JE, Grinspoon DH (1991) The faint young Sun problem. In: Sonett CP, Giampapa MS, Matthews MS (eds) The Sun in time. University Arizona, Tucson
20. Rye R, Kuo PH, Holland HD (1995) Atmospheric carbon dioxide concentrations before 2.2-billion years ago. Nature 378:603–605
21. Owen T, Cess RD, Ramanathan V (1979) Enhanced CO_2 greenhouse to compensate for reduced solar luminosity on early Earth. Nature 277:640–642. doi:10.1038/277640a0
22. Paris P von, Rauer H, Grenfell L, Patzer B, Hedelt P, Stracke B, Trautmann T, Schreier F (2008) Warming the early Earth—CO_2 reconsidered. Planet Space Sci 56:1244–1259. doi: 10.1016/j.pss.2008.04.008
23. Zahnle K, Schaefer L, Fegley B (2010) Earth's earliest atmospheres. Cold Spring Harb Perspect Biol. doi:10.1101/cshperspect.a004895
24. Rosing MT, Bird DK, Sleep NH, Bjerrum CJ (2010) No climate paradox under the faint early Sun. Nature 464:744–747. doi 10.1038/nature08955
25. Rosing MT, Bird DK, Sleep NH, Bjerrum CJ (2010) No climate paradox under the faint early Sun. Nature 464:744–747. doi 10.1038/nature08955
26. Minton DA, Malhotra R (2007) Assessing the massive young Sun hypothesis to solve the warm young Earth puzzle. Astrophys J 660:1700–1706. doi: 10.1086/514331
27. Shaviv NJ (2003) Towards a solution to the early faint Sun paradox: a lower cosmic ray flux from a stronger solar wind. J Geophys Res 108. doi: 10.1029/2003JA009997
28. Karoff C, Svensmark H (2010) How did the Sun affect the climate when life evolved on the Earth?—A case study on the young solar twin k^1 Ceti. arXiv:1003.6043v1 In press for Astron Nachr
29. Ribas I, de Mello GFP, Ferreira LD, Hebrard E, Selsis F, Catalan S, Garces A, do Nascimento JD Jr, de Medeiros JR (2010) Evolution of the solar activity over time and effects on planetary atmospheres. II. Kappa^1 Ceti, an analog of the Sun when life arose on Earth. Astrophys J 714:384–395. doi: 10.1088/0004-637X/714/1/384
30. Bahcall JN, Pinsonneault MH, Basu S (2001) Solar models: current epoch and time dependencies, neutrinos, and helioseismological properties. Astrophys J 555:990–1012
31. Chaplin WJ, Basu D (2008) Perspectives in solar helioseismology and the road ahead. Solar Phys 251:53–75. doi: 10.1007/s11207-008-9136-5
32. Manuel O, Friberg S (2003) Composition of the solar interior: information from isotope ratios. In: Lacoste H (ed) Local and global helioseismology, ESA SP-517
33. Krauss LM (1990) Correlation of solar neutrino modulation with solar cycle variation in p-mode acoustic spectra. Nature 348:403–407. doi 10.1038/348403a0
34. Boger J, Hahn RL, Rowley JK (2000) Do statistically significant correlations exist between the Homestake solar *neutrino* data and sunspots? Astrophys J 537:1080–1085
35. Ohmoto H, Watanabe Y, Kumazawa K (2004) Evidence from massive siderite beds for a CO_2-rich atmosphere before $\sim$ 1.8 billion years ago Nature 429:395–399
36. Kasting JF (2010) Faint young Sun redux. Nature 464:687–689

Chapter 4
Isotopes and Ice Cores

Abstract Isotopes produced by the interaction of galactic cosmic rays (GCRs) with atmospheric gases are incorporated in ice, sediments and plants. Beryllium-10 (^{10}Be) has a half life ($t_{1/2}$) of 1.5 Myr. Its short atmospheric residence time means that changes in ^{10}Be production can be traced at high resolution in stratified ice, but extracting the solar signal is complicated by climatic, geomagnetic and other factors and is performed confidently only for data from $\sim$ 100–1,000 yr BP, where the Maunder and other sunspot minima are clearly represented. Nevertheless progress in geomagnetic, palaeoclimatic and glaciological reconstruction should eventually unlock the solar data inherent in ^{10}Be sequences that already span the last 800 kyr. There is also the promise of identifying solar energetic particle (SEP) events from nitrate-rich layers in ice cores.

Keywords Galactic cosmic rays · Cosmogenic isotopes · Beryllium-10 (^{10}Be) · GRIP · EPICA · SEPs

The lunar and meteoritic record of solar history extends over several Gyr but, as we saw, it is extremely fragmentary. The narrative changes gear about 1 Gyr ago, when the discontinuous evidence of flares, CMEs and other explosive events on the Sun—sometimes lumped together as solar cosmic rays (SCRs)—can be supplemented by the evidence of isotopes produced by solar interaction with the galactic cosmic ray flux.

Here the role of the Sun is to shield the Earth's atmosphere so that it should be possible to derive changes in solar activity from isotopic levels. But the solar factor is not easily isolated, as GCRs are also deflected by the Earth's magnetic field, and climatic fluctuations may distort the evidence on the ground.

Moreover the measurement of stable and radioactive nuclides in meteorites suggests that the GCR flux has fluctuated by up to 10 % during the last 10 Myr, and, according to the long-lived isotope potassium-40 (^{40}K; $t_{1/2}$ 1.25 × 10^9 yr), by about 150 % during the last 0.5–1 Gyr [1]. These are of course long-term averages which may conceal even larger short-term oscillations [2]. Indeed, evidence has been found in the cosmogenic isotope [44]Ti from chondrites that fell in 1,810–1997 for a decline in GCR flux by 150–200 % over the last three centuries [3].

In other words our one constant is exceedingly variable. But the problem can be minimised, or at any rate identified, by comparison between isotopes with different origins coupled with fine tuning of the appropriate circulation models [4, 5].

Cosmogenic ^{10}Be

Cosmogenic isotopes are produced in the atmosphere by the interaction of primary and secondary GCR particles with nuclei of nitrogen, oxygen and argon.

Most GCRs have energies of ~ 100–10 GeV. There is a contribution by SCR, which are composed of ~ 98 % protons and 2 % heavier nuclei, but, with the exception of relatively rare SEPs, their energies are <100 MeV, and the resulting reactions are confined to the top of the atmosphere and to high latitudes [6]. At altitudes of 12–16 km the production rate is enhanced by secondary particles which, consisting mainly of neutrons, retain their energy more effectively than the protons of the primary rays [7].

Rocks may incorporate isotopes produced in the stratosphere as well as in the troposphere, where the cosmic ray flux is far lower because of atmospheric attenuation, but they generally lack the stratification required for constructing a time-sequence.

The cosmogenic isotope carbon-14 (^{14}C) can remain aloft for years as ^{14}CO$_2$ and it thus integrates conditions at a wide range of latitudes and altitudes before it is incorporated in plant tissues or in sediments as a carbonate. Beryllium-10 (^{10}Be) has a long $t_{1/2}$ of 1.5×10^6 yr, which means that its dating potential is in the region of at least 4 Myr. Once produced it is attached to aerosols and after 1–100 weeks it will be washed out and deposited in ice or other stratified deposits which can provide an independent chronology for the combined effects of production, transport and deposition.

Neutron monitors support a prima facie case for cosmic ray control of ^{10}Be production by the solar wind (Fig. 4.1) but their data, which began to be recorded in 1951, is too short to reveal any secular changes. The geomagnetic component is equally problematic, as there is substantial disagreement even for the last 50 kyr (for which ^{14}C dating provides age control) between reconstructions which exploit global data and those which trace the changes in field intensity at a single location (e.g. [8]). One can see why palaeomagnetic intensity data have been labelled an Achilles heel of solar activity reconstruction [9].

The shielding effect is at a maximum at low latitudes [10], and ^{10}Be production rates on the polar plateau in the last few hundred years have accordingly been little affected by geomagnetic field changes [4]. Thus it would seem sensible to limit the geomagnetic factor by sampling polar ice. But Central Greenland, for instance, receives much of its precipitation from lower latitudes, and some of the chemical constituents in the ice derive from eastern Asia [11], there have of course been

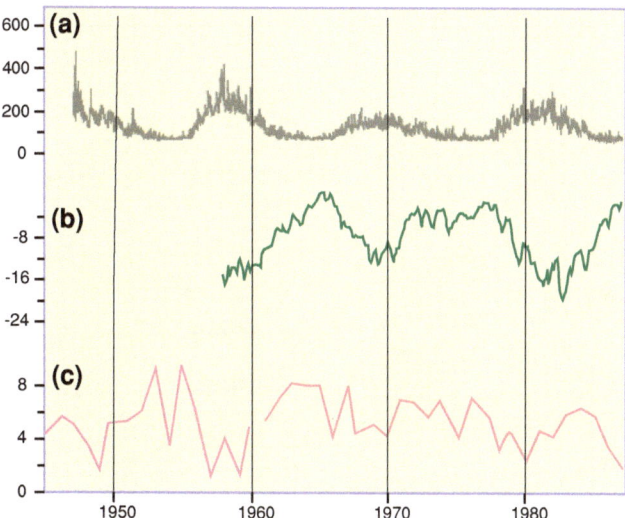

Fig. 4.1 Relationship between solar EUV, cosmic ray receipt and ^{10}Be flux. **a**: EUV proxy E $_{10.7}$ [36]; data from http://ftp.ngdc.noaa.gov/%20STP/SOLAR_DATA/SOLAR_UV/SOLAR2000// **b**: cosmic ray receipts from the Moscow Neutron Monitor at http://cr0.izmiran.rssi.ru/mosc/ main.htm; **c**: ^{10}Be flux for Dye3 core; datafrom [37]

substantial changes in global circulation during the Pleistocene [12], and ^{10}Be may linger a few years in the stratosphere after its production [13]. In short, ^{10}Be concentration (like ^{14}C) probably reflects the global mean production rate rather than purely local conditions [14].

Even so, ^{10}Be data for polar ice dating from 100 and 1,000 yr ago or perhaps earlier are thought to be driven primarily by the solar signal; at longer time scales palaeomagnetism plays a more important role [15]. For < 100 yr ago the local climate dominates the picture except during prolonged temperature minima, whence the poor correlation between ^{10}Be production and ^{10}Be concentration for 1939–2005 [16]. Atmospheric mixing and transport may thus mimic variations in solar activity [17]: the use of ^{10}Be to reconstruct solar history is evidently still 'in the development and testing stage' [7, p. 359].

Ice Cores

Two chronologically long ^{10}Be cores have been secured at high latitudes: the Greenland Ice Core Project (GRIP) ice core at Summit, which reached a depth of 3,170 m and extends from 386 kyr to 300 year BP [18], and EPICA Dome C core in Antarctica (EDC99), with ice at a depth of 3,190 m dated to ~800 kyr BP [19] (Fig. 4.2).

The primary control on the GRIP ^{10}Be data is considered to be precipitation rate combined with variations in ^{10}Be flux resulting from solar and magnetic

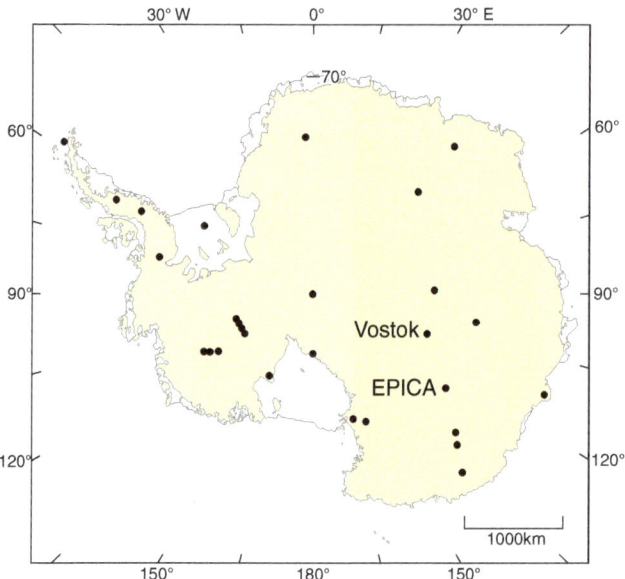

Fig. 4.2 Location of ice cores in Antarctica

modulation and possibly also variations in the cosmic ray flux. The EPICA sequence was anchored by a magnetic reversal at −3,160–3,170 m identified on the basis of enhanced [10]Be deposition, which was equated with the Matuyama-Brunhes geomagnetic reversal of 776 ± 12 kyr BP. The core displays numerous spikes in [10]Be with amplitudes up to an order of magnitude greater than adjacent samples. They were interpreted as local concentration effects rather than variations in production; the assumption that the main control has been geomagnetic is supported by broad agreement between a smoothed [10]Be flux curve and the virtual axial dipole moment (VADM) derived from marine sediments.

The solar factor appears to have been disregarded from the outset even though the possibility that the spikes represented 'enormous' flares was noted [20]. This suggestion is consistent with the results of spectral analysis of proxy temperature data [21] for the last 461 kyr, where the core is continuous and has a higher sampling rate than earlier portions: orbital fluctuations are seen to play a very minor role, with temperature excursions—including those corresponding to the 'spikes'—originating outside the climate system [22].

This conclusion does not eliminate geomagnetic mechanisms, although the only plausible candidate, the magnetic jerk, is associated more with changes in declination than with field intensity. Where the two potential explanations for the spikes converge is of course in geomagnetic storms (see Chap. 8) which illustrate how the distinction between solar and geomagnetic controls of isotope production is on occasion thoroughly blurred.

Moreover a climatic model is by no means the inevitable choice. Temperature for the EPICA Dome C core shows no clear relationship when plotted against [10]Be from

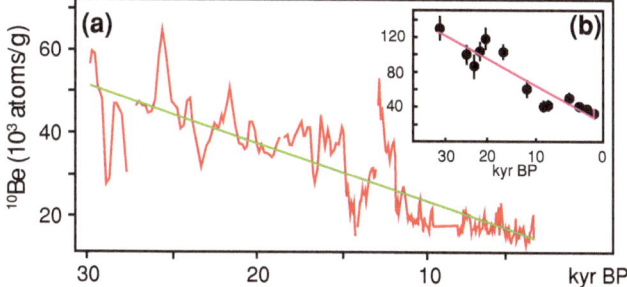

Fig. 4.3 a: [10]Be concentration in GISP2 (Greenland) core against age. Data ([10]Be (10^3 atoms/g) from [38] and at gcmd.nasa.gov/; $r^2 = 0.74$. There is a data gap at 1371-1510 m equivalent to 8017-9376 yr BP. **b**: [10]Be concentration in ice at Dome C (Antarctica) plotted from graphed data in [39]; y axis as for a; $r^2 = 0.87$

a SW Indian Ocean core (SK 200/23). The temperature record is consistent with the termination of the last glacial ~18–10 kyr BP whereas the minimum [10]Be concentration (4.52×10^8 atoms/g) was recorded at a downcore depth corresponding to ~30.34 kyr BP and the maximum value (2.67×10^9 atoms/g) at a downcore depth corresponding to ~21.59 kyr BP [23, 24]. The authors explain the mismatch by the effects of local bottom topography. It could equally well represent a real discrepancy stemming from independent solar control of [10]Be production

A similar thought is prompted by the major oscillations of the EPICA core. Among the many implications would be a change in the Sun's luminosity about 1 Myr ago which might help to account for the 'mid-Pleistocene revolution' that saw a shift from periodicities dominated by a 41 kyr frequency to the 100 kyr rhythm that now characterises terrestrial climate [24].

The cumulative decline in [10]Be concentration in the GISP2 core in Greenland and in the Dome C core in Antarctica (Fig. 4.3) provide further prima facie evidence for cumulative changes in solar activity, in this case an increase in the vigour of the solar wind. Such trends are not incompatible with orbital (Milankovitch) or other sources of change in irradiance, but, as discussed in the next chapter, they may provide the baseline for minor oscillations.

SEPs

We have encountered [21]Ne evidence from meteorites for higher flare incidence formerly. Flux values derived from ^{14}C and [81]Kr indicated higher flare activity in the past 10^4 or 10^5 years than indicated by [26]Al, [53]Mn and [3]He for the preceding few million years. The contrast may indicate an activity cycle with a period of 10^5–10^6 years or the occurrence of a few large flares during the last 10^5 years [25]. There is of course a third option: a cumulative change in solar luminosity.

The eruptive events that punctuate the polar ice cores cores provide further evidence of solar traces in the geological realm. Both flares and coronal mass ejections can release SEPs. Impulsive SEPs originate during rapid flares; gradual SEPs are associated with CMEs [26]. A continuous record of energetic particles in the heliosphere was obtained for nearly three solar cycles ([21–23], 1973–2001) by the Interplanetary Monitoring Platform-8, and it demonstrated a population of low energy (0.5–0.96 MeV) protons modulated by the solar cycle and supplemented by sporadic transient events.

Ice cores have substantially increased the SEP catalogue [27]. Ionization of the polar atmosphere by SEPs produces nitrates which are incorporated in the ice within 2–6 weeks. A high NO^{-3} ion concentration followed by a rapid rise in ^{10}Be content has been identified at a depth of 1,708.65 m in the GISP2 Greenland ice core and correlated with a date of 12,837 cal yr, which is marked by an abrupt rise in ^{14}C in the Cariaco (marine) varve chronology. The ^{14}C burst has been ascribed to a large SEP with a fluence of $\sim 1.3 \times 10^{11}$ protons/cm^2 [28].

For the period 1561–1950, nitrate-rich layers in Antarctic and Arctic cores identified 125 SEP events with a clear ~ 80–100 year Gleissberg periodicity and with six minima of which two were associated with the Maunder and the Dalton sunspot minimum [29]. A ~ 5 year periodicity was reported for nitrate abundance in Greenland ice for the period 1576–1991 and equated with the declining phase of the solar cycle [30] but as no such SEP pattern could be detected in a number of cycles, and, as the years of the Maunder Minimum displayed a ~ 20 year period, there was some support for the thesis of nitrate formation in the stratosphere by both galactic and cosmic rays [31].

The SEP model lost credibility from the finding that only one out of 14 well-resolved ice core records from Greenland and Antarctica displays a nitrate spike corresponding with the Carrington Event of 1859 [32] cf. [33]. The presence of ammonium, formate, black carbon and vanillic acid in the large spikes that have been analysed suggests they are better explained by biomass burning; hence the nitrate data cannot be used to evaluate SEP history. In any case, even when SEPs produce enhancement of NO in the stratosphere the signal will be blurred by other sources and may not reach reach ground level for 2 years [34]. Even so the issue of SEP history deserves further scrutiny not only in the context of flare frequency but also because stratospheric ionisation bears on current controversies over the role of GCRs in cloud formation [35]. And, if nothing else, the incident is a salutary reminder that periodicities can be spurious.

References

1. Wieler R, Beer J, Leya I (2011) The galactic cosmic ray intensity over the past 10^6–10^9 years as recorded by cosmogenic nuclides in meteorites and terrestrial samples. Space Sci Rev. doi: 10.1007/s11214-011-9769-9
2. Lifton NA, Bieber JW, Clem JM, Duldig ML, Evenson P, Humble JE, Pyle R (2005) Addressing solar modulation and long-term uncertainties in scaling secondary cosmic rays for in situ cosmogenic nuclide applications. Earth Planet Sci Lett 239:140–161

3. Usoskin IG, Solanki SK, Taricco C, Bhandari N, Kovaltsov GA (2006a) Long-term solar activity reconstructions: direct test by cosmogenic 44Ti in meteorites. Astron Astrophys 457:L25–L28
4. Usoskin IG (2008) A history of solar activity over millennia. www.livingreviews.org/lrsp-2008
5. Muscheler R, Beer J, Kubik PW (2004) Long-term solar variability and climate change based on radionuclide data from ice cores. In Pap JM, Fox P (eds) Solar variability and its effects on climate. AGU, Washington DC
6. Masaryk J, Beer J (2009) An updated simulation of particle fluxes and cosmognic nuclide production in the Earth's atmosphere. J Geophys Res 114, D11103. doi: 10.1029/2008JD010557
7. Beer J, Raisbeck GM, Yiou F (1991) Time variations of ^{10}Be and solar activity. In Sonett CP, Giampapa MS, Matthews MS (eds) The Sun in time. Univ Ariz Press, Tucson AZ
8. Gee J S, Cande SC, Hildebrand JA, Donnelly K, Parker RL (2000) Geomagnetic intensity variations over the past 780 kyr obtained from near-seafloor magnetic anomalies. Nature 408:827–832
9. Snowball I, Muscheler R (2007) Palaeomagnetic intensity data: an Achilles heel of solar activity reconstruction. Holo 17:851–859
10. Vonmoos M, Beer J, Muscheler R (2006) Large variation ns in Holocene solar activity: constraints from 10Be in the Greenland Ice Core project ice core. J Geophys Res 111, A10105, doi: 10.1029/2005JA011500
11. VanCuren RA, Cahill T, Burkhart J, Barnes D, Zhao Y, Perry K, Cliff S, McConnell J (2012) Aerosols and their sources at SummitGreenland—First results of continuous size- and time-resolved sampling. Atmos Env 52:82–97
12. Langen PL, Vinther BM (2009) response in atmospheric circulation and sources of Greenland precipitation to glacial boundary conditions. Clim Dyn 32:1035–1054
13. Beer J (2000) Long-term indirect indices of solar variability Space Sci Rev 984:53–66
14. Heikkilä U, Beer J, Abreu JA, Steinhilber F (2011) On the atmospheric transport and deposition of the cosmogenic radionuclides (^{10}Be): a review. Space Sci Rev, doi: 10.1007/s11214-011-9838-0
15. Usoskin IG, Horiuchi K, Solanki S, Kovaltsov GA, Bard ED (2009) On the common signal in different cosmogenic isotope data sets. J Geophys Res 114, A03112, doi: 10.1029/2008JA013888
16. Webber WR, Higbie PR (2010) A comparison of new calculations of ^{10}Be production in the Earths polar atmosphere by cosmic rays with ^{10}Be concentration measurements in polar ice cores between 1939–2005–A troubling lack of concordance: Paper #1. http://arxiv.org/abs/1003.4989
17. Lal D, Lingenfelter RE 1991 History of the sun during the past 4.5 Gyr as revealed by studies of energetic solar particles recorded in extraterrestrial and terrestrial samples. In Sonett CP, Giampapa MS, Matthews MS (eds) The Sun in time. Univ of Arizona Press, AZ
18. Yiou F et al (1997) ^{10}Be in the Greenland Ice Core Project ice core at Summit, Greenland. J Geophys Res 102:26,783–26,794
19. Raisbeck GM, Yiou F, Cattani O, Jouzel J (2006) ^{10}Be evidence for the Matuyama-Brunhes geomagnetic reversal in the EPICA Dome C ice core. Nature 444:82–84. doi: 10.1038/nature05266
20. Raisbeck GM, Yiou F, Cattani O, Jousel J (2005) "Spikes" of ^{10}Be in 700 Ky old ice from EPICA Dome C climatic or cosmic? Geophys Res Abs 7: 09466
21. Jouzel J and 31 others (2007) Orbital and millennial Antarctic climate variability over the past 800,000 years. Science 317:793–797
22. Reid JS (2010) A re-examination of ice age time series. At http://www.scienceheresy.com/2010_09/iceages/IceAgePaper.pdf
23. Khare N, Govil P, Kumar P, Mazumder A, Chopra S, Pattanaik JK, Balakrishnan S, Roonwal GS (2011) ^{10}Be as paleoclimatic tracer: initial results from south western Indian Ocean sediments. J Radioanal Nucl Chem 290:197–201

24. Head MJ, Gibbard PL (2005) Early-Middle Pleistocene transitions: an overview and recommendations for the defining boundary. Geol Soc Lond, Spec Pub 247:1–18, doi: 10.1144/GSL.SP.2005.247.01.01
25. Goswami JN, Jha R, Lal D (1981) Long Term Variations in Solar Flare Activity. J Astrophys Astr 2:201–212
26. Lario D, Simnett GM (2004) Solar energetic particle variation. In Pap JM, Fox P (erds) Solar variability and its effects on climate. AGU, DC
27. Usoskin IG, Solanki SK, Kovaltsov GA, Beer K, Kromer B (2006b) Solar proton events in cosmogenic isotope data. Geophys Res Lett 33, L08107, doi:10.1029/2006GL026059
28. LaViolette PA (2011) Evidence for a large flare cause of the Pleistocene mass extinction. Radiocarbon 53:303–323
29. McCracken KG, Dreschhoff GAM, Smart DF, Shea MA (2001) Solar cosmic ray events for the period 1561-1994: 2. The Gleissberg periodicity. J Geophys Res 106:21,599–21,609
30. Gladysheva OG, Kocharov GE, Kovaltsov GA, Usoskin IG (2002) Nitrate abundance in polar ice during the great solar activity minimum. Adv Space Sci 29:1707–1711
31. Kocharov GED, Ogurtsov MG, Dreschhoff GAM (1999) On the quasi-five year variation of nitrate abundance in polar ice and solar flare activity in the past. Solar Phys 188:187–190
32. Wolff EW, Bigler M, Curran MA, Dibb JE, Frey MM, Legrand MR, McConnell JR (2012) The Carrington event not observed in most ice core nitrate records. Geophys Res Lett 39:L08503, doi: 10.1029/2012GL051603
33. Schwenn R (2006) Space weather: the solar perspective. www.livingreviews.org/lrsp-2006-2
34. Wolff EW, Jones AE, Bauguitte S J-B, Salmon RA (2008) The interpretation of spikes and trends in concentration of nitrate in polar ice cores, based on evidence from snow and atmospheric measurements, Atmos. Chem Phys 8:5627–5634, doi:10.5194/acp-8-5627-2008
35. Mironova IA, Uosokin IG, Kovaltsov GA, Petelina SV (2012) Possible effect of extreme solar energetic particle event of 20 January 2005 on polar stratospheric aerosols: direct observational evidence. Atmos Chem Phys 12:769–778, doi:10.5194/acp-12-769-2012F
36. Tobiska WK, Woods T, Eparvier F, Viereck R, Floyd L, Bouwer D, Rottman G, White OR (2000) The SOLAR2000 empirical solar irradiance model and forecast tool. J Atmos Solar-Terr Phys 62:1233–1250
37. Aldahan A, Possnert G, Johnsen SJ, Clausen HB, Isaksson E, Karlen W, Hansson M, J. (1998) A 60 year long ^{10}Be record form Greenland and Antarctica. Earth Syst Sci 107:139–147
38. Finkel RC, Nishiizumi K (1997) Beryllium 10 concentrations in the Greenland Ice Sheet Project 2 ice core from 3–40 ka. J Geophys Res 102:26,699–26,706, doi:10.1029/97JC01282
39. Raisbeck GM, Yiou F, Fruneau M, Loiseaux JM, Lieuvin M, Ravel JC, Lorius C (1981) Cosmogenic ^{10}Be concentrations in Antarctic ice during the past 30,000 years. Nature 292:825–826, doi:10.1038/292825a0

Chapter 5
Cosmogenic Radiocarbon

Abstract The cosmogenic ^{14}C record for the Holocene is based on tree rings and marine deposits and now extends to 50,000 yr BP. Its significance for solar history is obscured by climatic and biological factors, but comparison with the ^{10}Be signal is helpful where the evidence is derived from ice cores; elsewhere the ^{14}C data can be supplemented by information on changing climatic zonation using cave and river deposits. Various periodicities have been identified by spectral analysis of the ^{14}C signal, including the ~ 2300 yr Hallstatt cycle and the 205 yr de Vries cycle. An apparent decline in atmospheric ^{14}C over the last 10 kyr extends the trend identified in the ^{10}Be signal for earlier millennia.

Keywords ^{14}C · Holocene · Hallstatt · De Vries · ITCZ · Grand minima · Grand maxima

The Holocene—the geological name for the last $\sim 10,000$ years—enjoys privileged status in the study of solar history because it is within comfortable reach of the ^{14}C technique, incorporates the 400 years of direct sunspot observation, and is endowed with a mass of archaeological and historical documents bearing on climate and hydrology. But the wealth of information has yet to yield significant insights into the solar interior, as the climatic and geomagnetic factors that hindered ^{10}Be analysis are just as obtrusive in the interpretation of radiocarbon sequences.

It has been shown, with the benefit of neutron monitor and sunspot data, that the ^{10}Be record for AD 1428–1950 embodies statistical, meteorological and SCR as well as GCR components and represents the time dependence of the GCR with a signal-to-noise ratio of ~ 8.1 [1]. It remains to be seen whether a similar assessment will eventually be possible for either ^{10}Be or ^{14}C data beyond the reach of instrumental or observational records.

Cosmogenic ^{14}C

Whereas ^{10}Be is produced by high energy spallation reactions of neutrons and protons with nitrogen and oxygen, ^{14}C is produced by thermal neutrons interacting with nitrogen [2]. One consequence of the difference is that ^{14}C production is more sensitive to changes in cosmic ray flux. On the other hand, whereas ^{10}Be (as we saw in Chap. 4) is soon attached to aerosols and washed out of the atmosphere by rain or snow, ^{14}C may be delayed for decades or centuries by its passage through the biosphere and the oceans before incorporation tree rings or carbonates, and the response of atmospheric ^{14}C level to solar/GCR changes will therefore be both damped and delayed: for example, a 200 years solar cycle with a production rate variation of 20 % will result in an atmospheric ^{14}C variation of 1 % delayed by $\sim$20 years [3]. In compensation the atmosphere will be well mixed (that is there will be no significant latitudinal effect) so that there is no merit in focusing on high latitude archives.

Despite these differences in 'system effects', principal component analysis of ^{14}C and ^{10}Be records which have been low-pass filtered (40 years) are dominated by a common signal—the production rate—which explains 76 % of the variance at multidecadal to multimillennial time scales [4]. Some put this figure as high as 90 % for low-pass filtered data (100 years) for cosmic ray flux over the last 10,000 years [2].

Unfortunately the signal is thought to embody both long-term, mainly geomagnetic effects and a short-term component due to solar modulation. The possibility that secular changes in atmospheric ^{14}C were primarily geomagnetic rather than solar in origin was being aired soon after the identification of ^{14}C. In 1956, Elsasser used measurements on ancient bricks dating from AD $\sim$200, 1465 and 1933, which showed a decline in magnetic moment from 0.71 to 0.456 G [5], to calculate the error in the ^{14}C age for an object 2000 years old as 240 years. Later, Bucha [6], on the basis of 300 archaeomagnetic measurements from Europe, Central America and Japan spanning 7000 years, concluded that there was an inverse relationship between fluctuations in radiocarbon production rate and changes in the Earth's magnetic moment.

This position is strongly supported by various tests. For example, the magnetic record of the last 80,000 years was used to create a synthetic ^{14}C atmospheric record (using a two-box model of the carbon cycle) for the period 40,000–14,000 yr BP which yielded ages in good agreement with U-Th dates [7]. A more complex analysis synthesised atmospheric ^{14}C data for the last 75,000 years, on the basis of geomagnetic field intensity, a reconstruction of North Atlantic bottom water circulation (on the grounds that the ocean contains 50 times more ^{14}C than the atmosphere) and two different box models, and it showed that calculated records of atmospheric ^{14}C were consistent with experimental determinations for the last 24 kyr [8]. As with ^{10}Be, therefore, the way is clear as regards the geomagnetic factor; but in addition allowance has to be made for the carbon cycle. Our grasp of both variables is progressing although, as it is driven primarily by the

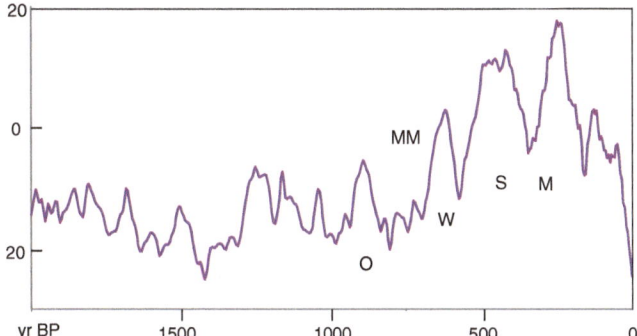

Fig. 5.1 Holocene ^{14}C fluctuations. Data from [37]. Minima: M = Maunder, S = Spörer, W = Wolf, O = Oort. MM = Mediaeval Maximum

needs of geophysics, climatology and biogeography rather than altruism, the data are not always immediately applicable to ^{14}C archives.

Grand Maxima and Minima

W. E. Maunder is best remembered for his endorsement of the observation by Gustav Spörer that there had been a period of several decades (1645–1715) when there were few or no sunspots. Indeed the anomaly came to be named after Maunder. The sunspot anomaly was reinforced by the record of aurora sightings and eventually by the radiocarbon signal preserved mainly in tree rings.

Early studies of the history of atmospheric ^{14}C had identified a number of fluctuations of which the greatest, named after its discoverer Hessel de Vries and amounting to 20 %, peaked in about 1690 [9]. It appeared to confirm the sunspot evidence for reduced solar activity during which (as explained in Chap. 4) the production of cosmogenic isotopes is enhanced [10]. The ^{14}C data revealed two other 'grand minima', the Spörer (AD 1420–1530) and the Wolf (AD 1280–1340), soon joined by the Oort (AD 1010–1050) and the Dalton (AD 1790–1820), and periods of anomalously low isotope production, notably the medieval grand (solar) maximum between about AD 1100 and 1250 (Fig. 5.1). They also indicated a number of periodicities of which the most pronounced have been found to be the ~ 88 years Gleissberg cycle and the Suess cycle of ~ 210 years and less securely the Hallstatt cycle of ~ 2300 years.

In their reconstruction of sunspot numbers using ^{14}C data Usoskin et al. [11] identified two different types of grand minimum: short (30–90 years) and long (>110 years) to a total of 27 minima for the last 11,000 years. The Maunder Minimum of AD 1645–1715 epitomises the first category. According to the Intergovernmental Panel on Climate Change (IPCC) [12], current evidence does not support the notion of a globally synchronous period of anomalous cold, although 'cold conditions appear… to have been considerably more pronounced in particular regions' and such regional variability 'can be understood in part as reflecting accompanying changes in atmospheric circulation'. But broadly synchronous phases are indeed reported from glaciers in Scandinavia, the Swiss Alps, and tropical glaciers in Peru [13]. And data from Antarctica and Greenland show that the intensity of meridional atmospheric circulation increased abruptly and synchronously in the polar South Pacific and the North Atlantic at about AD 1400 and may also have affected the circulation of middle and low latitudes [14]; summer temperatures were at their lowest for 500 years in both hemispheres during AD 1579–1730.

The links between solar history and terrestrial climate are undoubtedly circuitous. For instance, it has been suggested that the oceans thermohaline circulation, a large-scale system which is driven by density gradients and which includes sinking of warmer but more saline waters in the North Atlantic, with the Gulf Stream as a byproduct, slowed or even stopped following the injection of excess freshwater into the North Atlantic [15]. This trigger could represent melting induced by the Medieval Solar Maximum of AD 1100–1250 [16].

The flux of solar UV provides a plausible mechanism for generating climatic minima which are manifested over extensive regions. Reconstructions of spectral irradiance changes in the Sun which are based on facular brightening and reference to the range of Ca II emission in Sun-like stars indicate an increase of 0.7 % in the broad UV since 1675 [17]. Fluctuations in the UV region acting on the position of the Hadley cells are manifested in latitudinal shifts in the atmospheric circulation at middle and subtropical latitudes which in turn leave their imprint in terrestrial and marine sequences.

A related indicator of gross circulation geometry is the intertropical convergence zone (ITCZ)—sometimes termed the upward component of the Hadley cell—which shifts equatorward during decreased solar activity [18]. An analysis of the highly reflective cloud (HRC) satellite dataset revealed that the ITCZ in Africa and the Atlantic was displaced anomalously north by about 0.5° in the late 70 and early 80 s and anomalously south by a similar amount in the early 70 and mid 80 s [19]. The authors of the study suggest the effect may be related to long-period variability associated with the temperature and circulation of the Atlantic; it is worth noting that the northward shift is broadly coincident with the maximum of Cycle 21(June 1976–September 1986) and the southward shift with the minima before and after it. To be sure, the decennial shift is dwarfed by the annual cycle, which amounts to between 5° and 45° according to location, but any secular displacement in mean latitude can have a profound impact on rainfall at the arid margins of the ICTZ zone.

Among the phenomena on which this bears is the complex of oceanographic and meteorological effects of the El Niño/La Niña-Southern Oscillation (ENSO). Analysis of ENSO is generally based on coupled ocean-atmosphere models which assume that its causes are internal to the climate system [20–22] whereas the strength of the trade winds is both a symptom of Hadley Cell vigour and a major factor in the E-W displacement of warm surface waters that is at the root of the El

Niño cycle in the Pacific; a weakening of the trade winds raises the thermocline in the western Pacific and depresses it in the Indonesian realm. The intention here is not to throw out El Niño with the seawater but to suggest that at least part of the ENSO repertoire can be explained by deterministic (though not necessarily predictable) mechanisms arising from fluctuations in solar emission.

A modelling study which combined solar and orbital forcing has already yielded ENSO-like variance at centennial to millennial timescales [23], and analysis of mid-Holocene foraminifera from an eastern Equatorial Pacific core showed a reduction in variance by 50 % indicative of severe attenuation of ENSO amplitude at a time when the ITCZ was displaced northward [24]. More directly, sea-surface temperatures in the Tropical Pacific in the early Holocene display millennial-scale fluctuations which correlate with the cosmogenic record, with El Niño corresponding with solar minima [25]. In short, the ICTZ provides a link between ENSO processes and UV oscillations which may break the analytical obstacle created by the irregular length and duration of the constituent ENSO episodes.

Variations in the intensity of the Asian monsoon between 9600 and 6100 years ago have been traced by measuring past changes in the $\delta^{18}O$ of a stalagmite. Comparison with the ^{14}C record of tree rings for the same period [26] points to solar forcing although an analogous study in south China spanning the last 9000 years [27] suggests that orbital effects and changes in atmospheric and oceanic circulation also played a part in modifying the strength of the Asian monsoon.

The position and disposition of the jet streams in the upper troposphere and lower atmosphere provide additional measures of circulation shifts as they mark the frontiers between the major cells. The polar jet streams, at the poleward margin of the Hadley cell, allow such shifts to be detected because they govern the track and vigour of the mid-latitude cyclonic systems. Both storm track frequency [28] and the tracks themselves reveal shifts in latitude between solar maximum and minimum which are consistent with the models. The effect is especially clear in the North Atlantic, where solar maximum leads to a bunching and poleward shift in the tracks, and the Mediterranean basin, which is frequented by more depressions at solar minimum. An 11-year periodicity in the frequency of tropical cyclones in the Atlantic in 1871–1973 has long been viewed as a guide to the intensity of circulation over the North Atlantic [29].

Where cyclonic rainfall alternates with convectional rainfall it is sometimes possible to identify the switch from the behaviour of runoff, which can leave distinctive river deposits. Hence, by a very roundabout route, fluvial chronology can shed light on the former global circulation.

Two grand alluviation episodes are now well documented (1600–400 and 7500–4000 cal yr BP) and at least one more UV peak (or solar minimum) is indicated in the ^{14}C record for the last 25,000 cal yr BP [31]. It may be that against the solar timescale these minima will qualify as periodic.

Despite a lack of detailed chronologies in many parts of the subtropical realm some semblance of a regional palaeohydrological pattern for the two latest periods is beginning to emerge. In the Sahara-Sahel-Arabia region ($\sim$ 23–10°N), for example, a period of strong monsoons is dated to about 10.4–5.5 cal yr BP; desiccation in the eastern Sahara began $\sim$ 6.5–5.2 cal kyr at 22°N and $\sim$ 4.5–4.2 cal kyr at 17°N [32], a latitudinal effect consistent with the southward withdrawal of the monsoonal effect. In SE China (21°09′ N) weaker summer monsoons and stronger winter monsoons resulted from a shift in the ITCZ after 7.8 cal kyr BP, a change which is thought to have been synchronous over much of coastal SE Asia [33].

A latitudinal lag can be detected in Australia, where the onset of a drying phase dates from about 4.5–5.0 cal kyr in the wet tropics and $\sim$ 5.0 cal kyr in the sub-tropics [34]; in North Island, New Zealand, well dated sediment cores from floodplains, the continental slope and the continental shelf indicate that this happened (about 4000 yr BP) when the Circumpolar Westerly Vortex was intensified [35]. In the Atacama Desert of Chile (22–24°S) higher groundwater levels 8–3 kyr BP are explained by strengthening of the Pacific trade winds [36].

Secular Trends

The application of ^{14}C to the dating of organic material had been founded on the assumption that atmospheric levels had remained constant; the discovery of deviations from normality, both as a product of fuel burning since the mid nineteenth century and as an inescapable feature of the carbon cycle, led to the search for a calibration curve with which to correct ^{14}C dates to their calendric equivalent. The resulting plots were at first restricted to the last 12.4 cal kyr (the cal denoting that the age has been calibrated), the period for which dated tree-ring sequences could supply samples. Foraminifera and corals converted to the atmospheric equivalent with a site-specific marine reservoir correction later extended the sequence to the last 26 kyr; the present maximum attainable both by radiometric methods and by Accelerator Mass Spectrometry (AMS) is $\sim$ 60,000 years but background and contamination problems generally point to 50,000 years as a more realistic limit. Calibration to calendar ages is now possible for the entire period [37, 38].

The ^{14}C calibration curve shows that the atmospheric content of ^{14}C fell from $\sim$ 700 to $\sim$ 100 % between 26,000 and 11,000 yr BP; there was a further fall of over 200 % after 12,000 yr BP [39]. The general decline (Fig. 5.2a) has been ascribed to a slow change of the geomagnetic dipole [40], presumably an increase in the VADM to account for reduced cosmogenic production. Fluctuations in the ^{14}C curve are commonly detrended with the help of a sinusoidal curve, or by using a moving average, on the assumption that this allows the geomagnetic factor to be separated from the short-term (<200 years) solar signal. The procedure is not always made explicitly: long-term trends other than the 11-year variation were quietly removed by S. E. Forbush from widely used ground level ionization chamber readings for 1936–1968 and were thus deprived of secular significance [1].

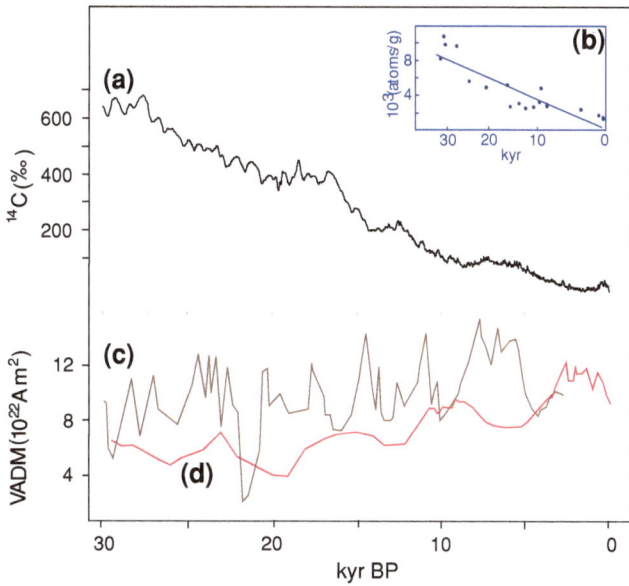

Fig. 5.2 a Atmospheric ^{14}C plot (‰); data from [39]; $r^2 = 0.97$. **b** In situ ^{14}C net concentration in ice crystals at Summit, Greenland; data from [41]; $r^2 = 0.85$; **c**: Geomagnetic field intensity from core SOH-4, Big Island, Hawaii, based on measurements on 100 lava flows dated by potassium/argon (K/Ar) and spanning the last 42 kyr [49]; **d**: plot based on GEOMAGIA50 database from the entire world, although Eurasia, the SW USA and western South America dominate the map as do determinations for the last 7000 yr, after [50]. VADM = virtual axial dipole moment

Figuer 5.2b is a plot of in situ ^{14}C production in GISP2 ice crystal samples [41]. Our geomagnetic field intensity sequences include one (Fig. 5.2c) based on terrestrial (lava) data but with a resolution reportedly as good as that of the best sedimentary records [42] and one (Fig. 5.2d) based on global data (50). The decline in ^{14}C is clear, although those responsible for the data in Fig. 5.2b (41) were more concerned with two epochs (9500–8500 and 27–32 kyr BP) when the production rate was twice as high as normal and one (12–16 kyr BP) when it was lower by a factor of ~ 1.5. In their view the in situ record in ice, which benefits from its location as regards geomagnetic distortion, is also immune from climate changes. The first two periods are represented by samples each spanning ~ 25 and 150–200 years respectively; the 16–12 kyr high is represented by only three ~ 25 year samples. The latter part of the record is less clear cut, as a ^{14}C decline over the last 1000 years is reported from the South Pole [43], whereas ^{10}Be and neutron monitor data suggest a steady increase in cosmic ray modulation since 1428 [1]. The relationship between cosmic ray flux and isotope generation evidently needs to be elaborated.

It has been suggested that solar irradiance is modulated by rotation of the radiative zone, or even of the solar core, rather than by the transit of active centres in the photosphere [44]. The notion is supported by the presence of the same period in Active Cavity Radiometer Irradiance Monitor (ACRIM) measurements [44] as well as by various indicators of rotational rigidity in the corona, a major source of solar UV radiation (see Chap. 7), although the transit of photospheric active regions doubtless makes a contribution to the net irradiance.

The next question must therefore be how far this might apply to fluctuations with decadal, centennial and millennial periodicities including those such as the Schwabe cycle and the Maunder Minimum that are conventionally associated with sunspot history. In this connexion we may note that analysis of the green coronal (53.03 nm) emission line for the period 1939–2005 reveals frequencies in the range 1.5–4 yr which originate mainly in the polar regions of the Sun and thus are not linked to the life cycle of active regions [45].

In spectral analysis of the neutrino data for the Home stake experiment for 1970–1990 one of the significant periodicities that emerged was 11 ± 1.5 yr [46]. Granted the limitations of the data, the finding suggests that an internal driver for the Schwabe cannot be ruled out. It would help to explain the persistence of an 11-year cycle during the sunspot-deficient Maunder Minimum [47].

For 10^2 and 10^3-year periods the analysis has to be more roundabout. One possibility is to compare the atmospheric ^{14}C sequence with total solar irradiance. The former, as we saw, portrays modulation of the GCR flux by the heliospheric magnetic field in the solar wind, and hence is strongly influenced by events in the corona. The variability of TSI is ascribed almost wholly to surface magnetic activity [48] but, according to the prevailing (Babcock) model [51], the sunspots and faculae that epitomise it are initiated by differential rotation in the body of the Sun.

In short, the suggestion here is that sunspots, faculae and other surface effects are secondary manifestations of the selfsame magnetic field that drives fluctuations in TSI, whereupon cosmogenic isotopes provide a more dependable guide to past and future solar periodicity than actual or synthetic sunspot records.

References

1. McCracken KG, Beer J (2007) Long-term changes in the cosmic ray intensity at Earth, 1428–2005. J Geophys Res 112:A10101. doi:10.1029/2006JA012117
2. Beer J, McCracken KG, Abreu J, Heikkilä U, Steinhilber F (2008) Long-term changes in cosmic rays derived from cosmogenic radionuclides. In: Proceedings of 30th International Cosmic Ray Conference 1:765–768
3. Muscheler R, Heikkilä U (2012) Constraints on long-term changes in solar activity from the range of variability of cosmogenic radionuclide records. Astrophys Space Sci Trans 7: 355–364. doi:10.5194/astra-7-355-2011
4. Abreu JA, Beer J, Steinhilber F, Christl M, Kubik PW (2011) ^{10}Be in ice cores and ^{14}C in tree rings: separation of production and climate effects. Space Sci Rev online. doi:10.1007/s11214-011-9864-y
5. Elsasser W, Ney EP, Winckler JR (1956) Cosmic ray intensity and geomagnetism. Nature 178:1226–1227
6. Bucha V (1969) Changes of the Earth's magnetic moment and radiocarbon dating. Nature 224:681–682

7. Mazaud A, Laj C, Bard E, Arnold M, Tric E (1991) Geomagnetic field control of ^{14}C production over the last 80 Ky: implications for the radiocarbon time-scale. Geophys Res Lett 18:1885–1888. doi:10.1029/91GL02285

8. Laj C, Kissel C, Mazaud A, Michel E, Muscheler R, Beer J (2002) Geomagnetic field intensity, north atlantic deep water circulation and atmospheric Δ^{14}C during the last 50 kyr. Earth Planet Sci Lett 200:177–190

9. De Vries HL (1958) Variation in concentration of radiocarbon with time and lLocation on Earth. In: Proceedings of Kon Ned Akad Wetensch B, 61:94–102

10. Eddy JA (1976) The Maunder Minimum. Science 192:1189–1202

11. Usoskin IG, Solanki SK, Kovaltsov GA (2007) Grand minima and maxima of solar activity: new observational constraints. Astron Astrophys 471:301–309

12. IPCC (Intergovernmental Panel on Climate Change) (2001) Climate change 2001:the Scientific Basis. WMO/UNEP, Geneva

13. Licciardi JM, Schaefer JM, Taggart JR, Lund DC (2009) Holocene glacier fluctuations in the Peruvian Andes indicate northern climate linkages. Science 325:1677–1679

14. Kreutz KJ, Mayewski PA, Meeker LD, Twickler MS, Whitlow SI, Pittalwala I I (1997) Bipolar changes in atmospheric circulation during the Little Ice Age. Science 277:1294–1296

15. Broecker WS (2000) Was a change in thermohaline circulation responsible for the Little Ice Age? Proc Nat Acad Sci 97:1339–1342

16. Jirikowic JL, Damon PE (2005) The medieval solar activity maximum. Clim Change 26: 309–316

17. Lean J (2000) Evolution of the Sun's spectral irradiance since the Maunder Minimum. Geophys Res Lett 27:2425–2428

18. Versteegh GJM (2005) Solar forcing of climate. 2: evidence from the past. Space Sci Rev 120:243–286

19. Waliser DE, Gautier C (1993) A satellite-derived climatology of the ITCZ. J Clim 6: 2162–2174

20. Bjerknes J (1969) Atmospheric teleconnections from the equatorial pacific. Monthly Weather Rev 97:163–172

21. Cobb KM, Charles CD, Cheng H, Edwards RL (2003) El Niño/Southern oscillation and tropical pacific climate during the last millennium. Nature 424:271–276

22. Chen D, Cane MA, Kaplan A, Zebiak SE, Huang D (2004) Predictability of El Niño over the past 148 years. Nature 428:733–736

23. Emile-Geay J, Cane MA, Seager R, Almasi P (2007) El Niño as a mediator of the solar influence on climate. Paleoceanography 22:doi:10.1029/2006PA001304

24. Koutavas A, deMenocal PB, Olive GC, Lynch-Stieglitz J (2006) Holocene El Niño-Southern Oscillation (ENSO) attenuation revealed by individual foraminifera in eastern tropical Pacific sediments. Geology 34:993–996

25. Marchitto TM, Muscheler R, Orti JD, Carriquiry JD, van Geen A (2010) Dynamical response of the tropical Pacific Ocean to solar forcing during the Holocene. Science 330:1378–1381

26. Neff U, Burns SJ, Mangini A, Mudelsee M, Fleitmann D, Matter A (2001) Strong coherence between solar variability and the monsoon in Oman between 9 and 6 kyr ago. Nature 411:290–293

27. Wang Y, Cheng H, Edwards RL, He Y, Kong X, An Z, Wu J, Kelly MJ, Dykoski CA, Li X (2005) The Holocene Asian monsoon: links to solar changes and North Atlantic climate. Science 308:854–857

28. Vita-Finzi C (2008) Fluvial solar signals. Geol Soc London Spec Pub 296:105–115

29. Colebrook JM (1976) Trends in the climate of the North Atlantic Ocean over the past century. Nature 263:576–577

30. Mayewski PA, Maasch KA, Yan Y, Kang S, Meyerson EA, Sneed SB, Kaspari SD, Dixon DA, Osterberg EC, Morgan VI, van Ommen T, Curran MAJ (2005) Solar forcing of the polar atmosphere. Ann Glaciol 41:147–153

31. Vita-Finzi C (2010) Alluvial history and climate crises. Spec Trans Am Philos Soc 1: 115–124

32. Hoelzmann P, Gasse F, Dupont LM, Salzmann U, Staubwasser M, Leuschner DC, Sirocko F (2004) Palaeoenvironmental changes in the arid and subarid-belt (Sahara-Sahel-Arabian Peninsula) from 150 ka to present. Dev Paleoenv Res 6:219–256

33. Yancheva G, Nowaczyk NR, Mingram J, Dulski P, Schettler G, Negendank JFW, Liu J, Sigman DM, Peterson LC, Haug GH (2007) Influence of the intertropical convergence zone on the East Asian monsoon. Nature 445:74–77

34. Donders TH, Haberle SG, Hope G, Wagner F, Visscher H (2007) Pollen evidence for the transition of the eastern Australian climate system from the post-glacial to the present-day ENSO mode. Quat Sci Rev 26:1621–1637

35. Gomez B, Carter L, Trustrum NA, Palmer AS, Roberts AP (2004) El Niño–Southern Oscillation signal associated with middle Holocene climate change in intercorrelated terrestrial and marine sediment cores, North Island, New Zealand. Geology 32:653–656

36. Betancourt JL, Latorre C, Rech JA, Quade J, Rylander KA (2000) A 22,000-year record of monsoonal precipitation from northern Chile's Atacama Desert. Science 289:1542–1546

37. Reimer PJ et al (2009) IntCal09 and Marine09 radiocarbon age calibration curves, 0–50,000 years cal BP. Radiocarbon 51:1111–1150

38. Hoffmann DL, Beck JW, Richards DA, Smart PL, Singarayer JS, Ketchmark T, Hawkesworth CJ (2010) Towards radiocarbon calibration beyond 28 ka using speleothems from the Bahamas. Earth Planet Sci Lett 289:1–10

39. Reimer PJ et al (2004) IntCal04 terrestrial radiocarbon age calibration, 0–26 cal kyr BP. Radiocarbon 46:1029–1058

40. Bard E (1998) Geochemical and geophysical implications of the radiocarbon calibration. Geochim Cosmochim Acta 62:2025–2038

41. Lal D, Jull AJT, Pollard D, Vacher L (2005) Evidence for large century time-scale changes in solar activity in the past 32 Kyr, based on in-situ cosmogenic ^{14}C in ice at Summit, Greenland. Earth Planet Sci Lett 234:335–349

42. Garnier F, Laj C, Herrero-Bervera E, Kissel C, Thomas D (1996) Preliminary determinations of geomagnetic field intensity for the last 400 kyr from the Hawaii Scientific Drilling Project core, Big Island, Hawaii. J Geophys Res 101. doi:10.1029/95JB03844

43. Lal D (2009) Radiocarbon concentrations in South Pole ice samples of ages 120–954 yr. At http://www.usap-data.org/entry/NSF-ANT05-38683/

44. Sturrock PA (2009) Combined analysis of solar neutrino and solar irradiance data: further evidence for variability of the solar neutrino flux and its implications concerning the solar core. Solar Phys 254:227–239

45. Vecchio A, Carbone V. 2009. Spatio-temporal analysis of solar activity: main periodicities and period length variations. Astron Astrophys 502:981–987

46. Liritzis I (1995) Quasi-periodic variation in the solar-neutrino flux revisited. Solar Phys 161:29–47

47. Letfus VI (2000) Sunspot and auroral activity during *Maunder* Minimum. Solar Physics 197:203–213

48. Kopp G, Lean JL (2011) A new, lower value of total solar irradiance: evidence and climate significance. Geophys Res Lett 38. doi:10.1029/2010GL045777

49. Garnier F, Laj C, Herrero-Bervera E, Kissel C, Thomas D (1996) Preliminary determinations of geomagnetic field intensity for the last 400 kyr from the Hawaii Scientific Drilling Project core, Big Island, Hawaii. J Geophys Res 101. doi: 10.1029/95JB03844

50. Knudsen MF, Riisager P, Donadini F, Snowball I, Muscheler R, Korhonen K, Pesonen LJ (2008) Variations in the geomagnetic dipole moment during the Holocene and the past 50 kyr. Earth Planet Sci Lett 272:319–329

51. Babcock HW (1961) The topology of the Sun's magnetic field and the 22-year cycle. Astrophys. J 133:572–587

Chapter 6
The Solar Cycle

Abstract The $\sim$11-year solar cycle was first identified in the fluctuation of the number and location of sunspots and, soon after, their magnetic polarity and the dipole component of the Sun's magnetism as a whole. The cycle is manifested in satellite records of total and variable solar luminosity, in the open solar magnetic flux, which can be estimated from solar observation or from the aa geomagnetic index, and in the Sun's radio emission The incidence of flares and coronal mass ejections is linked to the cycle in complex fashion. The Babcock model of sunspot generation relates the cycle to twisting of magnetic field lines by differential solar rotation; deeper links to internal dynamics are revealed by neutrino data. Periodicities which were not obvious when solar observation was limited to the photosphere may turn out to be more significant as regards the solar interior than the Schwabe or Hale cycle.

Keywords Hale cycle · TSI · Babcock model · Solar minimum · Spörer · Wolf · $E_{10.7}$

The $\sim$11-year sunspot cycle, first identified, as we saw, by S. H. Schwabe, was formalised by Rudolf Wolf [1]. Much effort has since been expended on studies of its varying length, amplitude and spectral constitution and on linking its fluctuation with climatic changes; and the enterprise has acquired special value in the study of solar development now that helioseismology allows sunspots to be viewed in three dimensions and dedicated satellites can trace sunspot development continuously.

This is not to minimise the remarkable achievements of the early observers. Galileo firmly established sunspots as constituents of the solar atmosphere and used them to demonstrate the Sun's rotation, and Scheiner noted the pattern of their migration that was to blossom into the butterfly diagram of Edward Maunder. The next major step was taken by George Ellery Hale [2], who identified the $\sim$22-year magnetic polarity cycle of sunspot pairs, and Horace Babcock [3], who schematised the cycle in terms of lines of force drawn out and twisted by the Sun's differential rotation.

C. Vita-Finzi, *Solar History*, SpringerBriefs in Astronomy,
DOI: 10.1007/978-94-007-4295-6_6, © The Author(s) 2013

The Sunspot Cycle

Much has since been learnt about the temperature and magnetism of sunspots, the bright areas that adjoin them, and intervening portions of the photosphere. Throughout, sunspot number has been viewed in one way or another as a measure of solar activity. The assumption that high sunspot values denote a hotter Sun at first rested on simple experience. It was implicit in attempts to link sunspots to agricultural success, as William Herschel did with the corn prices of 1801, and later more directly to various components of climate. To equate 'solar activity' (a term now restricted by some to surface magnetic processes) with solar luminosity was to jump the gun, as the mechanism linking the two was not known, although it seemed plausible that changes in the sun's convection could lead to changes in luminosity [4].

The belief was to be confirmed in due course by satellite measurements of the solar constant, although the difference between solar maximum and solar minimum turned out to be a mere 0.1 %.

In 1848 Wolf introduced a formula for standardising sunspot records, R = k (10g + f), where g is the number of sunspot groups visible on the solar disc, f the number of individual sunspots, and k a correction factor which varies between observing stations. The last value was devised in order to normalise the data to what Wolf would have reported, and it is recalculated every year. Wolf was able to trace sunspot totals back to the solar cycle of 1755–1766 which came to be known as Cycle 1.

In 1858 Richard Carrington reported that the spots experienced an equatorward drift from latitude 40° (modern authors would say 30°) at solar minimum to latitude 5° at sunspot maximum; he also concluded that lower latitudes rotate faster than higher latitudes, an effect which came to be named after Gustav Spörer. In 1889 Spörer drew attention to the apparent absence of spots in the seventeenth century, a phenomenon confirmed and elaborated by W.A. Maunder in 1890. In 1904/1913 Maunder plotted the progressive migration of sunspots in the two hemispheres to produce what is now known as a butterfly diagram (Fig. 6.1). The spectroheliograph, invented by Hale in 1892, allowed observation of the Sun in different wavelengths, and it revealed, among other things, the association of sunspots with bright plages and shortlived flares, an assemblage that came to be termed an active region.

Though all these observations were fundamental they were initially skin deep. The crucial link with the solar interior was forged by 1908 with Hale's discovery that sunspots were magnetic phenomena (with fields measuring up to 1,500 gauss (G)), as this demonstrated that the Sun's visible activity was a manifestation of its magnetic field and thus of its internal dynamics. Hale also showed that, when the sunspots were present as pairs, the spot leading in the direction of the Sun's rotation was opposite in polarity to its follower, with a mirror image of this pattern in the other hemisphere, and that the arrangement was reversed at the next cycle. As the Sun's entire magnetic field reversed its polarity at each cycle, a complete

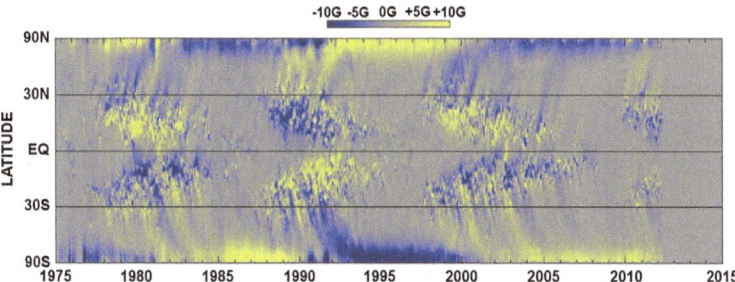

Fig. 6.1 Map of Sun's magnetic field over three solar cycles. *Yellow*: magnetic field directed out of the Sun; *blue*: field directed into the Sun; *high latitude streaks*: poleward meridional flow carrying magnetic remains of sunspots. After NASA/ MSFC/David Hathaway

(Hale) cycle measured a total of ~ 22 years. These observations opened the door to an interpretation of solar processes couched in the language of magnetohydrodynamics, the dynamics of fluids (including plasmas) which are electrically conducting.

Regrettably, although active regions are potentially more effective guides to solar activity than sunspots alone, their detailed history is even shorter than that of sunspots as they are better traced on solar magnetograms than on images of the photosphere. Thus sunspots retain their historical value as tokens of cyclicity spanning four centuries even if, as shown below, their absence is not necessarily proof of solar inactivity.

Solar Luminosity

From the outset sunspots and their attendant features were linked to climate and weather. The musings of Herschel on harvests have gained some credence from evidence that solar UV influences climatic zonation as a whole rather than solely temperature. There is of course ample evidence that the solar cycle is manifested in solar luminosity [5]. Indeed, satellite observations have shown that, as faculae often occupy a larger area than the associated sunspots, they may create peaks in the TSI both before and after the dips caused by sunspot blocking. But it is also clear that TSI may be influenced by various other factors operating at timescales of days, hours and minutes, and varying widely in significance, which include the temperature of the photosphere, the solar radius, the solar granulations and the p-mode oscillations in the Sun's deep interior [6].

The Earth Radiation Budget (ERB) experiment on the Nimbus-7 satellite ran from 1978 to 1993, the first high precision irradiance study. Many other experiments have followed but, as the accuracy of their calibration is put at ± 0.2 %, cumulative changes may well be obscured. A particular problem arises from the 2-year gap (July 1989–October 1991) between the measurements by the Activity

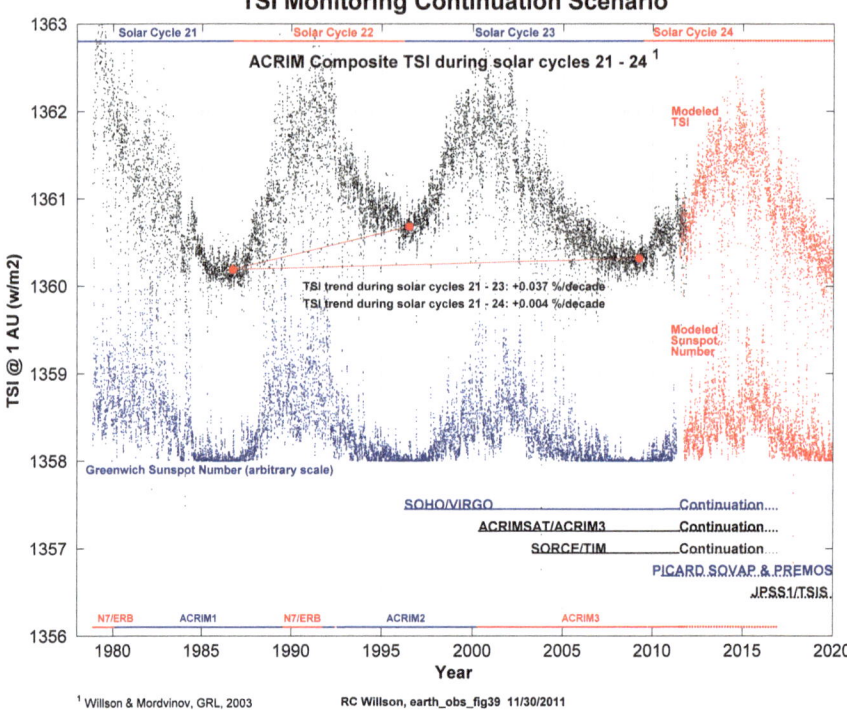

Fig. 6.2 Total Solar Irradiance and sunspot number 1978–2011 and modelled plots to 2020. Note TSI minimum trends for cycles 21–23 (+ 0.037%/decade) and 21–24 (+ 0.004%/decade). Courtesy of R. C. Willson [33]

Cavity Radiometer Irradiance Monitoring (ACRIM) I experiment on the Solar Maximum Mission (SMM) satellite, and ACRIM II, on the Upper Atmosphere Research Satellite (UARS). If the gap is bridged by reference to ERB data, there is an apparent increase in irradiance of c. 0.08 Wm^{-2} (0.03 %) between the minimum at the end of Cycle 21 and the minimum at the end of Cycle 22 [7] (Fig. 6.2). Others dispute this inference: a version originating in the Physikalisch-Meteorologisches Observatorium Davos (PMOD), which uses a different procedure for bridging the ACRIM gap, shows no significant upward trend [6].

Analysis of the cycle itself, however, has much benefited from the technical advances of recent decades both in solar monitoring and in related fields. Perhaps the greatest beneficiary has been our understanding of how the solar cycle is manifested in different wavelengths (spectral solar irradiance or SSI as opposed to TSI), as it bears on solar modulation of the Earth's atmosphere. On the whole the shorter wavelengths, which contribute least to the luminosity, fluctuate most markedly in the course of the solar cycle, and they may have disproportionate effects (Fig. 6.3). Some 8.7 % of the solar flux energy comes from wavelengths of 30–400 nm; the UV wavelengths (10–400 nm) contribute 17–60 % of the TSI

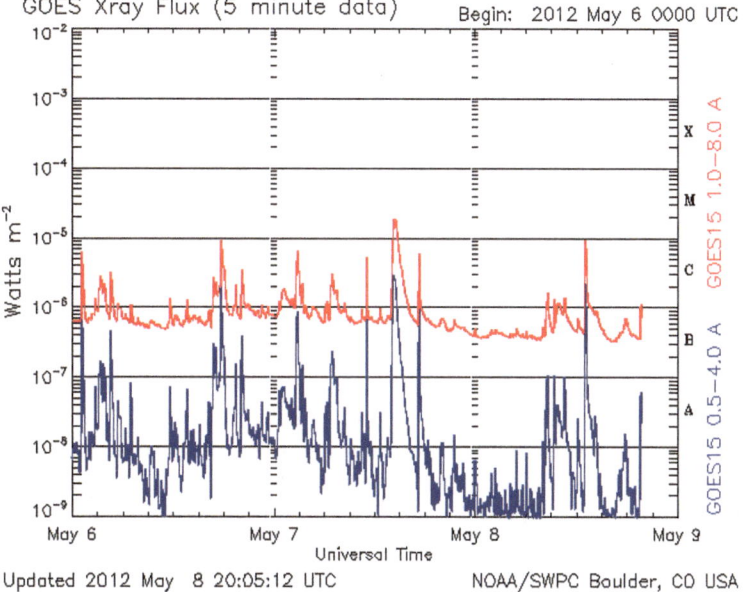

Fig. 6.3 Solar X-ray flux (5 minute data, two wavebands) observed by GOES-15 satellite, the sixteenth in the Geostationary Operational Environmental Satellite (GOES) system operated by the US National Oceanic and Atmospheric Administration. It had instruments to monitor the magnetosphere, cosmic background radiation and charged particles as well as flares

variation [7]. Above 290 nm the variation is below experimental uncertainties. Those below 60 nm may vary by over 300 % over the cycle, and 100 % change in the flux at ∼100 nm accounts for changes of some 1,000 K in the exosphere (500–1,000 km above the surface).

The Magnetic Connection

The Hale period had been identified in drought data for the SW USA by 1971 and in tree-ring records of that region by 1979. Deuterium/hydrogen (D/H) ratios on bristlecone pines spanning 1000 years indicated a 22.36-year. But the lack of a plausible link between solar magnetism and the terrestrial climate led Dicke [8] to propose that the requisite variation in solar luminosity had to be driven by a source deep in the Sun rather than by surface magnetism. In 1989, a 21.8-year periodicity that had dominated marine air temperature trends between 1856 and 1986 was linked to the solar magnetic cycle [9] and a 20–22 year period identified with the Hale cycle in Midwest USA climate data [10]; Dicke's objection remains—but only if solar cyclicity is recorded on Earth by climatic signals alone. As shown in

Chap. 7 the neutrino evidence provides an alternative terrestrial measure of the cycle.

Of course the Hale cycle is important in its own right whatever its significance for the workings of the inner Sun. The classic explanation for the very existence of sunspots combines differential rotation with the notion of magnetic field lines. In the qualitative model proposed by Horace Babcock [3], as the solar equator rotates faster than the polar regions the poloidal field lines at low latitudes in the course of 5–8 years are twisted into a toroidal pattern. During this growth phase of the cycle the leader of the two spots will have the same polarity as the hemisphere which hosts it. Surface flows carry magnetic fields towards the pole with opposite polarity, that is to say south-pointing flux to the north magnetic pole and vice versa, thus weakening the dipole field until it reverses.

If part of the toroidal field becomes unstable a section of it may rise buoyantly to the surface and break through it to form a pair of sunspots surrounded by an active region. For this buoyant process to bring up a field of sufficient strength the time required is put at less than a year, the strength at about 10^5 G, and the location at around latitude $60°$, where the increase in azimuthal field strength is greatest owing to differential rotation [11]. The active region will gradually decay in the course of ~ 3 years in response to degradation of the magnetic flux tubes by supergranule motion around them.

Granted the value of sunspot patterns for tracing the evolution of the ~ 11-year cycle, the fact remains that its periodicity persisted during sunspot minima. The downcore concentration of the cosmogenic isotope ^{10}Be at Dye 3 in Greenland reflects vigorous Schwabe cyclicity during the Maunder [12]. Similarly, frequency analysis of the ^{14}C content of Japanese cedar trees for AD 1557–1629, a period which includes the Spörer as well as the Maunder, shows that the ~ 11 years period was manifested throughout [13].

Indicators of solar activity other than sunspots have been described as 'more physical' [14], presumably in the sense that they are the direct product of internal processes rather than thermal blemishes on the photosphere resulting from the interplay between differential rotation and magnetic flux lines. Three such guides to the solar cycle are the 10.7 cm radio emission, the Ca II K line, and the He 1,083 nm line.

The $F_{10.7}$ index has been measured daily (in Canada) since 1947, and ranges from <50 to >300 units (one solar flux unit is equivalent to 10^{-22} W m^{-2} Hz^{-1}). It represents the solar magnetic flux emerging through the photosphere at a wavelength of 10.7 cm (2,800 MHz) and originates mainly as plasma trapped over active areas [15]. Used in the past as an input to ionosphere models it now provides estimates of the effect of the ionosphere on GPS frequencies. Its simplicity—instrumental readings in place of somewhat subjective assessment of the sunspot number—has led to its adoption in place of the sunspot number for many applications. There is considerable scatter on a plot of sunspot number against the 10.7 cm flux, but the fact that the latter never drops below a value well above zero (~ 67) even during solar minimum [16] suggests that it does indeed provide the sounder basis for ionospheric research.

Besides its convenience and reasonable agreement with sunspot data, the 10.7 cm value also tallies with measurements of the EUV (10–120 nm) solar radiation, which strongly influences the atmosphere (especially its density) above 200 km. Comparison between the observed and modelled orbital decay of 7 satellites showed that the index is not an accurate index of total EUV over the solar cycle [17], but closely reflects that component of EUV radiation from active regions, and is thus a convincing substitute for sunspot numbers.

Even so there is room for improvement in its representation, and a EUV proxy, $E_{10.7}$, has been devised for use in climate research and engineering applications as well as ionosphere research [18]. Expressed in the same radio units as $F_{10.7}$, it denotes the EUV energy deposited at the top of the Earth's atmosphere at 1 AU. An important difference is that $F_{10.7}$ represents coronal emissions from the corona and the transition region [19] whereas $E_{10.7}$ integrates chromospheric as well as coronal EUV emissions; from the viewpoint of solar physics, rather than climatology, it is surely useful to have a means of distinguishing between the two sources. Moreover, $F_{10.7}$ is measured once a day whereas $E_{10.7}$ has a resolution measured in minutes. Consequently, although $F_{10.7}$ responds to solar flares in the form of solar noise bursts, known as *tenflares* if they exceed 100 % of the background noise level, $E_{10.7}$ can document the rise and decay of large solar flares.

A second, alternative route to EUV analysis is by assessment of the structures primarily responsible for UV emission: active regions, enhanced network (into which an active region decays soon after a plage has formed), and the quiet Sun, using Ca II K spectroheliograms. The Ca II K line (393.3 nm) is appropriate because its centre is manifested in the chromosphere.

Analysis of 1,400 spectroheliograms covering the maximum of solar cycle 21 and much of cycle 22 (1980–1996) showed that plages and enhanced network covered about 13 and 10 % respectively of the solar disk during solar maximum periods and zero at solar minimum. But the chromospheric (or active) network into which the enhanced network disperses (and which outlines the supergranule cells) may continue to cover much of the disk at solar minimum [20]. UV emission at solar minimum is thus more diffuse than at solar maximum and poorly expressed by sunspots and plages alone.

A third alternative guide to the solar cycle is the He 1,083 absorption line, which is linked to the chromosphere and the corona. The He line bears on the solar wind. The helium abundance (expressed as the ratio He/H) is the most variable quantity in the solar wind as it ranges from <4.5 to 30 %. Although the record is evidently limited to the period during which appropriate devices were in space, there are several reports of a strong variation between solar maximum and minimum since 1974 (Fig. 6.4). Some of them find that abundance is linked to solar wind speed especially during solar minimum, where the speed dependence may be four times stronger than at solar maximum [21]. Moreover, there are indications that, again at solar minimum, the fast solar wind (~ 750 km/s), with a photospheric composition, originates mainly in the areas of open magnetic field termed coronal holes at high solar latitudes, and its slow counterpart (~ 400 km/s), with a coronal composition, from the streamer belt along the solar middle latitudes [22].

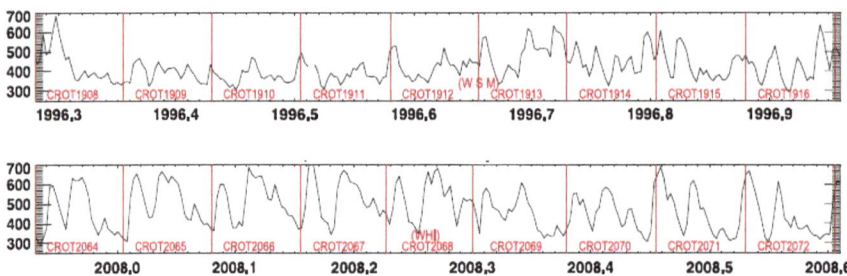

Fig. 6.4 Time series of solar wind velocity for nine 27-day solar rotations during solar minimum for (upper) 10 August–8 September 1996 and (lower) 20 March–16 April 2008, when sunspot numbers were at their lowest for 75 years. Four days have been added to the solar rotation start times to allow for travel to Earth. After [23]

The crucial point here is that, whatever the validity of any such scheme, variations in the strength of the solar wind may owe more to heliospheric latitude than to position in the solar cycle, a notion endorsed by observations made during the solar minimum of 2008. That minimum was exceptionally quiet, with sunspot numbers at the lowest level in 75 years and solar wind magnetic field strength lower than previously recorded, but large low-latitude coronal holes persisted and gave rise to strong, long and repeated high-speed solar wind streams. Among the effects was the incidence of outer radiation belt fluxes 3–4 times higher than during the previous (1996) solar minimum [23].

Related Effects

It has long been surmised that the incidence of flares and CMEs peaks at solar maximum. Like many tidy cyclicities this has proved an oversimplification, to the detriment of space weather forecasting (Fig. 6.4). To be sure, there is a preponderance of flare events near the cycle maximum, put by some authorities at several a day compared to fewer than one a week at solar minimum, but some of the largest events have occurred near solar minimum. They include two major flares on 5 and 6 December 2006, which at their peak created 20,000 times more radio emission than the rest of the Sun; and a flare on 28 October 2003 which measured X28 where the usual range for X flares is 1–9. Similarly, CMEs average one every 5 days at solar minimum and perhaps 3 a day at solar maximum, but a fast CME on 1 August 2010 followed the longest solar minimum for more than a century.

'Potentially the most powerful way of finding observational evidence for past variations in the solar output is to consider their influence on the planets, and on the earth in particular' [24, p. 468]. Analysis of 45,000 observations made in AD 1450–1948 showed a low incidence of aurorae in the Dalton, Maunder and Spörer [25] but to identify fossil CMEs we need at the very least evidence that they extended to anomalously low latitudes.

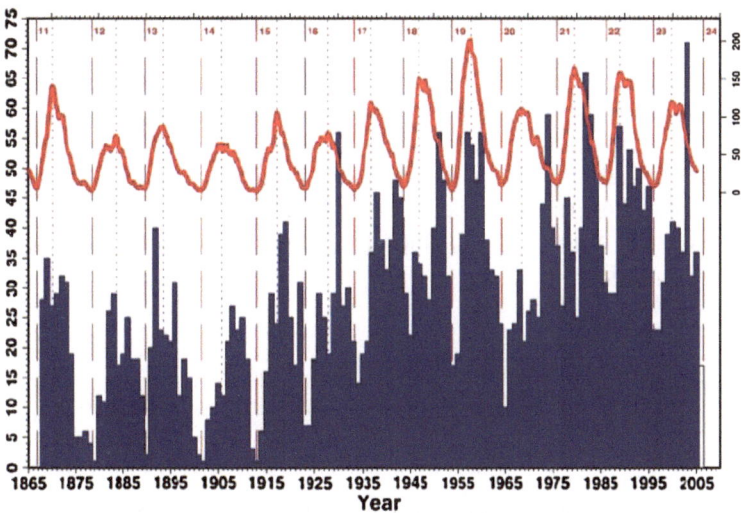

Fig. 6.5 Relationship between number of magnetic storms in year (*vertical bars*) and the solar cycle (*smoothed curve*). (courtesy of S. Macmillan/British Geological Survey (Natural Environment Research Council))

The geomagnetic record is far shorter but more explicit. The solar flare observed by Carrington in 1859 was followed 17 h later by widespread disruption of telegraphic systems as well as auroral displays. A magnetic storm, a term introduced by Alexander von Humboldt, is a temporary disturbance of the Earth's magnetosphere and, apart from rare eyewitness reports, it will have gone unrecorded until the introduction of the geomagnetic magnetogram in about 1848. (The 1859 event is well documented.) Here too there is misconception that the solar cycle is in control, but severe storms can occur in almost any month although there is some increase in the years around solar maximum and during the declining years of the cycle. And the most severe storm on record, though associated with a great spot group [26], occurred on 18 September 1941, 4 years 5 months after solar maximum.

Other, more geocentric indicators of solar activity include the aa index, which was introduced in 1868 and is based on eight 3 h magnetometer readings at two roughly antipodal stations (at present Hartland in the UK and Canberra in Australia) standardised for latitude. It includes data from only one longitude in each hemisphere but the annual averages agree with the genuinely global index Am to within 1 % [27]. The maximum yearly averages fall tidily, sometimes with two or more peaks, within the 14 solar cycles between 1844 and 1997 [28] (Fig. 6.5).

The aa index also reveals secular change: data for 1868–1996 show an upward drift during the twentieth century indicating an increase in the total magnetic field derived from the Sun by a factor of 1.4 since 1964 and a factor of 2.3 since 1901 [29] cf [30]. The finding is further evidence that the instantaneous level of activity, such as the sunspot number, provides a partial picture of the interplanetary field as, unlike the ^{10}Be record, it disregards secular changes in the background [31].

On the other hand both the aa and the ^{10}Be records are compromised by their geomagnetic context in that their signals are distorted to an unknown extent by terrestrial factors, and it is fortunate that their signals can be evaluated, though only over 19 years, by direct measurement from the Ulysses spacecraft. The outcome is that the open flux varies over the solar cycle by a factor of 2 and that, in combination, the centennial (i.e. since 1905) and cyclic variations amount to a factor of ~ 4.5 [32].

References

1. Eddy   J   (1976)   The   Maunder   minimum.   Science   192:1189–1202.   doi:10.1126/science.192.4245.1189
2. Hale GE (1908) On the probable existence of a magnetic field in Sun-spots. Astrophys J 28:315–343
3. Babcock HW (1961) The topology of the Sun's magnetic field and the 22-year cycle. Astrophys J 133:572–587
4. Siquig RA, Hoyt DV (1980) Sunspot structure and the climate of the last one hundred years. In: Pepin RO, Eddy JA, Merrill RB (eds) The ancient Sun. Pergamon, New York
5. Woodard MF, Libbrecht KG (2003) Spatial and temporal variation in the solar brightness. Solar Phys 212:51–64
6. Pap JM, Frölich C (1999) Total solar irradiance variations. J Atmos Solar-Terr Phys 61:15–24
7. Floyd L, McMullin D (2008) Long-term variations in UV and EUV solar spectral irradiance. International workshop solar variability, Earth's climate and the space environment. Bozeman MT
8. Dicke RH (1979) Solar luminosity and the sunspot cycle. Nature 280:24–27
9. Newell M, Newell RE, Hsiung J, Zhongxiang W (1989) Global marine temperature variation and the solar magnetic cycle. Geophys Res Lett 16:311–314. doi:10.1029/GLO16i004p00311
10. Chen H-L, Rao AR (1998) Periuodicity in midwest US climatic data: the Hale cycle signal or the luni-solar nsignal? Stoch Hydrol Hydraul 12:205–220
11. Spruit HC (2011) Theories of the solar cycle: a critical view. In: Miralles MP, Sánchez Almeida J (eds) The Sun, the solar wind, and the heliosphere. IAGA spec sopron book ser 4. doi:10.1007/978-90-481-9787-3_5
12. Beer J, Tobias S, Weiss N (1998) An active Sun throughout the Maunder minimum. Solar Phys 181:237–249
13. Miyahara H, Masuda K, Nagaya K, Kuwama K, Muraki Y, Nakamura T (2007) Variation of solar activity from the spoerer to the Maunder minima indicated by radiocarbon content in tree-rings. Adv Space Res 40:1060–1063
14. Thomas JH, Weiss NO (2008) Sunspots and starspots. Cambridge University Press, Cambridge
15. Tapping KF (1987) Recent solar radio astronomy at centimetre wavelengths the temporal variability of the 10.7 cm flux. J Geophys Res 92:829–838. doi:10.1029/JD092iD01p00829
16. IPS (2012) Ionosphere preediction service at www. ips.gov.au
17. Anderson AD (1964) On the inexactness of the 10.7 cm flux from the Sun as an index of the total extreme ultraviolet radiation. J Atmos Sci 21:1–14
18. Tobiska WK, Woods T, Eparvier F, Viereck R, Floyd L, Bouwer D, Rottman G, White OR (2000) The Solar000 empirical solar iraadiance model and forecast tool. J Atmos Solar-Terr Phys 62:1233–1250

19. Lean J (1987) Solar ultraviolet irradiance variations: a review. J Geophys Res 92:839–868

20. Worden JR, White OR, Woods TN (1998) Spectroheliograms: implications for solar ultraviolet irradiance variability. Astrophys J 496:998–1014

21. Aellig MR, Lazarus AJ, Steinberg JT (2001) The solar wind helium abundance: variation with wind speed and the solar cycle. Geophys Res Lett 28:2767–2770

22. Feldman U, Landi E, Schwadron NA (2005) On the sources of fast and slow solar wind. J Geophys Res 110:AO7109. doi:10.1029/2004JA010918

23. Gibson SE, Kozyra JU, de Toma G, Emery BA, Onsager T, Thompson BJ (2009) If the Sun is so quiet, why is the Earth ringing? A comparison of two solar minimum intervals. J Geophys Res 114:A09105. doi:10.1029/2009JA014342

24. Gough DO (1977) Theoretical predictions of variations in the solar puitput. In: White OR (ed) The solar output and its variation. Colorado Ass University Press, Boulder

25. Silverman SW (1992) Secular variation of the aurora for the past 500 years. Rev Geophys 30:333–351

26. Newton HW (1941) The great spot group and magnetic storm of September 1941. Obs 64: 161–165

27. Lockwood M, Whiter D, Hancock B, Henwood R, Ulich T, Linthe HJ, Clarke E, Clilverd MA (2006) The long-term drift in geomagnetic activity: calibration of the aa index using data from a variety of magnetometer stations. Ann Geophys 24: 3411–3419

28. ftp.ngdc (National Geophysical Data Center).noaa.gov/stp/ geomagnetic_data/aastar/

29. Lockwood M, Stamper R, Wild MN (1999) A doubling of the Sun's coronal magnetic field during the last 100 years. Nature 399: 437–439

30. Clilverd MA, Clarke E, Ulich T, Linthe J, Rishbeth H (2005) Reconstructing the long-term aa index. J Geophys Res 110:A07205. doi: 10.1029/2004JA010762

31. Solanki SK, Schüssler M, Fligge M (2000) Evolution of the Sun's large-scale magnetic field since the Maunder minimum. Nature 408:445–447

32. Lockwood M, Owens M (2009) The accuracy of using the *Ulysses* result of the spatial invariance of the radial heliospheric field to compute the open solar flux. Astrophys J 701: 964, doi: 10.1088/0004-637X/701/2/964

33. Willson RC, Kwan S, Hellzon R, Scafetta N (2011) LASP/TRF diagnostic test and results for the ACRIM3 experiment

Chapter 7
Solar Rotation

Abstract In 1630 Scheiner observed that the Sun's period of rotation varied with latitude. Differential solar rotation is an integral component of the Babcock model of sunspot generation, first proposed in 1961, but the mechanism remains qualitative. Cogent explanations for the latitude effect had to await the advances in computing of the twentieth century but the resulting models were eventually invalidated by the findings of helioseismology. The passage of active zones across the solar disk is generally thought to account for the mean ~ 27-day periodicity in the Sun's irradiance, but as TSI has a period of 32 d it appears to lag behind the sunspot signal. An alternative explanation, for which there is support from periodicities in neutrino flux, is that, as once proposed by Dicke, the TSI is driven primarily by an internal 'chronometer'.

Keywords Sunspot migration · EUV · Babcock model · Dicke cycle · Neutrino flux · Helioseismology

The possibility that solar activity might reflect the Sun's inertial motion, and this in turn the dynamics of the planetary system as a whole, has long been mooted (e.g. [1]). In view of the large mass contrasts—Jupiter's mass is 1/1047 that of the Sun—the notion appears implausible, but calculations show that in the long-term the combined effect of its planetary retinue could effect some minor modulation of the Sun's internal motion and hence the solar dynamo [2].

Some of the numerous periods that have been identified by spectral or other statistical analysis of TSI [3] may turn out to reflect gravitational effects. At present the ~ 11 and ~ 22-year periodicities in luminosity and magnetism that were considered in Chap. 6 are confidently ascribed to processes operating within the Sun but there is little agreement over an autonomous mechanism for them and external triggers or stimuli cannot be ruled out.

Rotation

Solar rotation (Ω) is a key property of the Sun as it is may be influenced by large-scale internal circulation, the impact of the solar wind and processes in the heliosphere [4]. To be sure, as sunspots (and plages) are linked by magnetic field lines to deeper layers, their rotation rate is not necessarily representative of the photosphere [5]. But they bring a history measured in centuries which benefits from graphic documentation.

The average rate of solar rotation, as we saw, was probably substantially faster than now when the Sun first joined the Main Sequence, so that some continuing, irregular deceleration is to be expected from the braking effect of the solar wind. In 1980 daily measurements using the Doppler effect did not reveal changes greater than 20 m s^{-1}, that is to say no trend exceeding 1 % [6], although, as noted below, a general acceleration during the declining phase of solar cycle 20 was reported on the basis of Doppler data [7]. By 1996 the rms error had been reduced to 7 m s^{-1} [4], but the period of observation remains too short for any unambiguous trend to be detected.

An added twist to the evidence comes from the possibility that variations in rotation rate are themselves subject to secular change. For example short-term periodicities in the equatorial rotation rate in 1986–1995 (cycle 22) are apparently absent in cycle 23, and they do not necessarily stem from differences in data quality [8]. Especially tantalising is a 1.3-year periodicity which has been detected in the rotation rate near the base of the convection zone during 1995–1999 but not after 2001 and which is also reported in sunspot activity, solar-wind speed and the interplanetary magnetic field [8], as it promises to illuminate the links between the solar interior and the heliosphere.

The earliest determinations were of course made by reference to sunspots. Once they had been recognised as features of the solar atmosphere their displacement, recognised by Johannes in 1611, promptly led to the identification of a period of rotation. Galileo put it at 29.5 d; Scheiner proposed 27–28 d. The mean sidereal rate for 1612 established from 199 drawings by Thomas Harriot was 13.3°/day (27.07 d), 6 % slower than at the present day [9]. The observations by Johannes Hevelius in 1642–1644 similarly led to the conclusion that the equatorial synodic rotation was 3–4 % higher than the present [10] but this was contradicted by a later analysis [11]. The 53 years of observation by J Picard from 1666 to 1719 appear to show that equatorial rotation was slower during the Maunder Minimum [12] [13], but the Greenwich photoheliograph daily observations for 1888–1964, spanning solar cycles 13–19, indicate a linear deceleration at the solar equator totalling 1 % during a period when solar activity was generally on the increase [14], the converse of what was reported for the seventeenth century [15] and during the declining phase of cycle 20 [7].

The early observers soon detected differential rotation. Scheiner was perhaps the first to report, in 1630, that the rate was higher near the poles. Detail was promptly added to his observation by Richard Carrington (1863), Spörer (1874) and Maunder (1905). The present synodic period is 26.2 at the equator and 39.3 near the poles [16].

This effect too appears to have fluctuated over the last four centuries. Picard's observations appear to indicate that during 1666–1719 there was a higher level of differential rotation with latitude [12]; and during the period just before the Maunder Minimum the rate of change in rotation between 0 and $\sim 20°$ was enhanced by 3x [10].

Quantitative analysis of the differential rotation had to await the second half of the twentieth century, when adequate computing power became available [17]. Interpretation of the rotation took into consideration motion in the convective zone and the requirements of dynamo models; but when helioseismology made it possible to investigate rotation within the convection zone, 30 years 'of theoretical efforts were nullified by the severe law of observations' [17, p 276].

Viewing the outcome in a more positive light, however, analysis of the solar oscillations data obtained by the Michelson Doppler imager on SOHO showed that the variation of Ω with latitude observed at the surface extends through much of the convection zone and that, in contrast, beneath the tachocline the rotation rate is nearly constant. (The MDI data also showed the high latitude rotation rate decreased between 1995 and 1999 and then started increasing [18]. However, the requisite simulations call for fully developed turbulence in fluids with low Prandtl number (Pr, the ratio of kinematic viscosity to thermal diffusivity and thus a measure of convection relative to conduction) at very high Reynolds numbers (Re, a measure of turbulence) where earlier simulations assumed almost laminar or slightly turbulent convection [19].

Luminosity

Variations in luminosity over the 27 day period, as already noted, are conventionally ascribed to the passage of active areas across the solar disk [20]. The blocking effect of sunspots is countered by bright features, notably faculae and network elements, and TSI consequently experiences a peak at Schwabe maximum. But, whereas the sunspot activity period is ~ 27 d [21], the quasi-rotation period for TSI is ~ 32 d and may vary from one cycle to another [3]. One suggestion for the conflict is that TSI lags behind sunspot activity by 29 d [3].

Although blocking by sunspots reduces the Sun's radiative output photospheric faculae and chromospheric plages and network enhance visible and UV emission respectively but there is still uncertainty over whether an additional factor is required to account for the Schwabe irradiance cycle: one calculation using sunspots and faculae alone underestimated irradiance changes over the 11-year period by a factor of about 2 [22]. An explanation was sought in the neglected active magnetic network and, alternatively, in global photospheric temperature; the identification of magnetic sources shown by Ca K imaging to be brighter than the background Sun during the declining phase of cycle 22 suggested that they satisfactorily accounted for both the bolometric (that is, luminosity at all wavelengths) and the UV irradiance cycles [22].

But the need remains to account for the mismatch in the timing of the TSI and sunspot cycles noted above. It will be recalled that Dicke [23] proposed that the main control of solar luminosity was 'a chronometer' inside the Sun, a suggestion which came from inspection of the sunspot cycle and its persistence during the Maunder Minimum. Dicke's preference was for an oscillator located at the base of the Sun's convective zone and he suggested that the associated magnetic field floated to the surface in 10–15 years. The variant advanced here—the Dicke cycle—is that a quasi-periodic UV spike, originating in the radiative zone, is distorted at the surface by differential rotation [24].

A case built on sunspot data has been dismissed by those who argue that 300 years of dependable sunspot record are not sufficient to reveal (or rule out) any phase memory, the notion that, rather than a random walk in phase (that is to say a phase which drifts randomly), the solar cycle manifests phase excursions with respect to a stable oscillator. An alternative view is that there is a subtle memory effect—phase and amplitude are strongly correlated over the 26 half-cycles since 1,700—which can be explained by mean field theory without the help of Dicke's chronometer [25]. The correlation between phase and amplitude, so the argument runs, generates a slow random walk equivalent to a finite memory extending over several hundred years, and great solar anomalies are not anomalies but simply large fluctuations.

Focusing on the UV wavelengths has a number of advantages. As well as avoiding the minefield of sunspot statistics the data bear on a wide range of chemical and physical processes in the Earth's middle and upper atmosphere. As UV luminosity varies substantially over the solar cycle [26] it provides unambiguous fluctuations in the signal. The requisite data are provided by the $F_{10.7}$ proxy $E_{10.7}$, which lends itself to analyses spanning several cycles. In order to bring out any underlying periodicities such data are sometimes subjected to a filter width of 81 d, that is to say three synodic solar rotations, but in the present context a two-year moving average is preferable because it may obscure subtle periodicities but will not prejudge them. A pilot analysis examined the period 1992–2003 to encompass a complete cycle and then focused on the year 1993 on the grounds that it appeared unremarkable.

The presence of multiple peaks meant that locating the baseline was an arbitrary matter, and they yielded periods ranging from 23 to 33 d, with 26.6 the value for maxima alone. Ambiguity in the luminosity record is to be expected if it is the product of the passage of active regions. Sunspots normally first appear at about 30°, where the rotation period is ~ 29 d, and gradually migrate to lower latitudes. Moreover, to judge from a century of solar Ca II K (393.37 nm) data, plages account for much of the UV variability at wavelengths <240 nm [27]. Faculae—and thus plages—are more widely distributed than sunspots: in 2003–2005 they extended to latitude 60° N and S, where the synodic rotation is ~ 34 d [16], whereas sunspots were confined to the zone bounded by 30° N and 40° S. In other

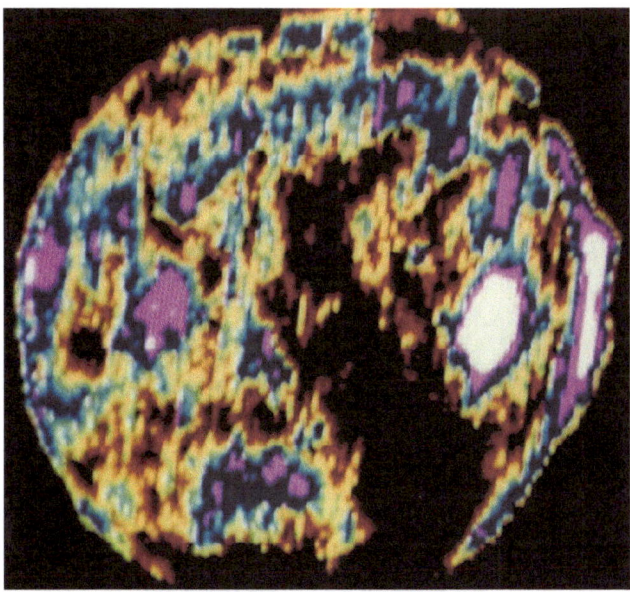

Fig. 7.1 Coronal hole imaged in extreme UV (EUV) by Skylab astronauts (courtesy of NASA)

words one cannot expect crisp oscillation in UV flux when the latitudinal effect is added to the evolution of active regions.

As it happens the supposed rotation effect appears to be absent from UV wavelengths >290–340 nm [28] and perhaps >210 nm [29]. Moreover the TSI measured by the ACRIM (1980–1982) and ERB (1978–1982) radiometers did not display a ~28 day peak. Such a peak was recorded at 205 nm [20] but the effect should not be confined to a narrow band given that the wavelengths affected by enhanced emission from active regions range from soft X-rays (0.1–10 nm) to decimetric radio waves [16]. The UV effect is manifested in the density of the thermosphere, in response to the 27 day rhythm of the solar EUV signal [30], in stratospheric ozone content [31] [32], and thence in the intensity of the 30 hPa meridional circulation in winter at middle and high latitudes [33]. Besides their climatic significance these effects are potentially routes to verifying astronomical inferences about solar behaviour.

A ~27 day recurrence also applies to coronal holes [16], and which are most prominent just before solar minimum (Fig. 7.1). The period varies little with latitude [34], further evidence for quasi-rigid rotation of the coronal plasma and for a magnetic link between some of the corona's large-scale magnetic features and processes deep within the Sun [35]. The period is manifested in the galactic cosmic ray flux [36], presumably in response to variations in the strength of the protective solar wind. The role of GCRs in cloud formation remains uncertain, but a direct effect on the troposphere is indicated by the ~27 day periodicity in cloud cover that was observed during solar maximum years during the interval 1980–2003 over

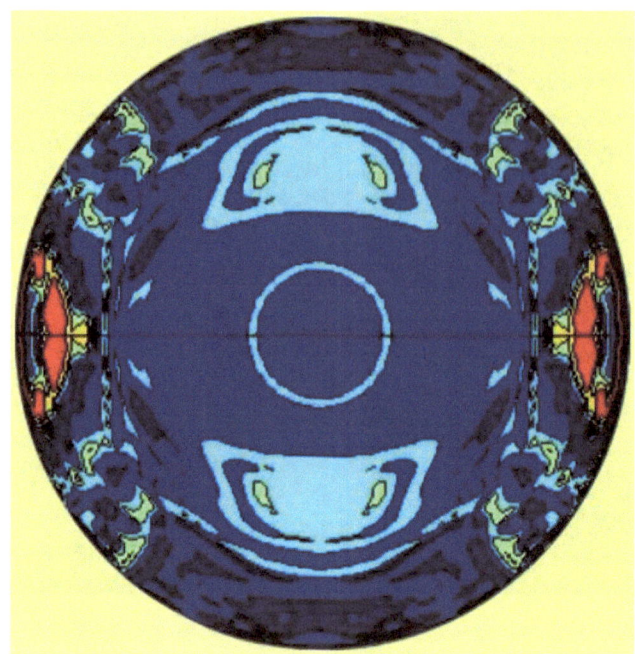

Fig. 7.2 Data from the GALLEX/GNO neutrino detector indicate solar neutrino flux variations that match the rotation rates of the areas shown in red on this cross-sectional map of the solar interior rotation (courtesy of M. Weber & P. Sturrock (Stanford), J. Scargle (NASA/ARC), SOHO/MDI, GALLEX/GNO)

the western Pacific (using outgoing long wave radiation as proxy), as the $F_{10.7}$ flux displays a similar periodicity with an amplitude 30 times greater than at solar minimum [37].

The Neutrino Flux

That the Dicke chronometer may lurk in the radiative zone is consistent with the finding of helioseismology that the rotation rate of this zone is latitudinally constant [38]. At the tachocline the sidereal rotation rate jumps from 26.55 d at $r = 0.66$ to 25.14 d at $r = 0.74$. Above the tachocline, however, the rate varies with latitude and radius from a minimum of ~ 42.08 d at $r = 1$ at the poles to a maximum of ~ 24.61 d at $r = 0.93$ at the equator [39].

Modulation of the flux could be the result of either density variations or magnetic field variations, and the latter seemed the more promising possibility 'since the Sun can accommodate much greater inhomogeneities in its magnetic field than in its density' [40].

Besides its apparent rotational rigidity the radiative zone appeared to offer two more advantages for the study of neutrino modulation: it is larger, and it contains gas that, being at higher pressure, could contain stronger magnetic field [39].

Power spectrum analysis of data collected over 24 yr at the detector in Dakota yielded evidence of a 26.31 day cycle. It elicited the comment that, if real, the periodicity 'could be the synodic value of the rotation rate of the Sun's radiative zone' [40, p. 413]. The search was extended to the gallium experiments on the grounds that the shorter half life of the gallium capture product ^{71}Ge (11.43 d) renders it better suited to a search for variations on a timescale in weeks than the chlorine capture product ^{37}Ar ($t_{1/2}$ 35.0 d). For the GALLEX-GNO data the biggest peak was at 26.88 d, matching the rate obtained by helioseismology in the convection zone at ~ 0.85 r, and comparative analysis with SOHO/MDI data showed no evidence of modulation in the radiative zone [39] (Fig. 7.2).

In order to circumvent the problem of very low count rates the Homestake and GALLEX datasets were combined, omitting 13 % of the Homestake sequence to avoid overlap, to yield (on the basis of a procedure termed joint power statistic) as the top peak in the frequency band 0–20 yr^{-1} a period of 30.82 d. Solar irradiance data obtained by the SMM/ACRIM radiometer for the same period gave an identical result. The conclusion reached was that irradiance as well as neutrino flux is modulated by processes in a solar core rotating with a period of 30.82 d, that is to say a sidereal frequency of 12.85 yr^{-1} [41]. This may well be the period of the Dicke cycle which was put forward to account for variations in the TSI that had previously been ascribed to sunspot migration across the solar disk.

Analysis of the data obtained by the Super-Kamiokande (water) neutrino detector had given the markedly faster rate of 13.75 d for the inner radiative zone [42], which raised the possibility of a second tachocline between the core and the radiative zone and thus of an inner dynamo which could generate a strong internal magnetic field and, more important in the present context, a second solar cycle [41]. There is evidence that the period is manifested at the surface in the sunspot blocking function [42, 43], whereas no correlation could be found with the maximum and minima of solar cycle 23 [44], further evidence that the link between sunspots and irradiance is a devious one.

References

1. Fairbridge RW, Shirley JH (1987) Prolonged minima and the 179-yr cycle of the solar inertial motion. Solar Phys 110:191–220
2. Callebaut DK, deJager C, Dulhau S (2012) The influence of planetary attractions on the solar tachocline. J Atmos Sol-Terr Phys 80:73–78. doi: 10.1016/jastp.2012.03.005
3. Li KJ, Xu JC, Liu XH, Gao PX, Zhan LS (2010) Periodicity of total solar irradiance. Solar Phys 267:295–303
4. Ulrich RK, Bertello L (1996) Solar rotation measurements at Mount Wilson over the period 1990–1995. Astrophys J 465:L65–L68
5. Howard R (1984) Solar rotation. Ann Rev Astron Astrophys 22:131–155

6. Scherrer PH, Wilcox JM (1980) Doppler observations of solar rotation. Astrophys J 239:L89–L90
7. Howard R (1976) A possible variation of the solar rotation with the activity cycle. Astrophys J 210:L159–L161
8. Javaraiah J, Ulrich RK, Bertello L, Boyden JE (2009) Search for short-term periodicities in the Sun's surface rotation: a revisit. Solar Phys 257:61–69. doi: 10.1007/s11207-009-9342-9
9. Herr RB (1978) Solar rotation determined from Thomas Harriot's sunspot observations of 1611 to 1613. Science 202:1079–1081. doi: 10.1126/science.202.4372.1079
10. Eddy JA, Gilman PA, Trotter DE (1976) Solar rotation during the maunder minimum. Sol Phys 46:3–14
11. Abarbanell C, Wöhl H (1981) Solar rotation velocity as determined from sunspot drawings of J. Hevelius in the 17th century. Solar Phys 70:197–293
12. Ribes JC, Nesme-Ribes E (1993) The solar sunspot cycle in the Maunder minimum AD 1645 to AD 1715. Astron Astrophys 276:549–563
13. Casas R, Vaquero JM, Vazquez M (2006) Solar rotation in the 17th century. Sol Phys 234:379–392. doi: 10.1007/s11207-006-0036-2
14. Eddy JA, Noyes RW, Wolbach JG, Boornazian AA (1978) Secular changes in solar rotation, 1888–1964. Bull Am Astr Soc 10:400–401
15. Eddy JA, Gilman PA, Trotter DE (1977) Anomalous solar rotation in the early 17th century. Science 198:824–829
16. Phillips KJH (1992) Guide to the Sun. Cambridge University Press, Cambridge
17. Paternò L (2010) The solar differential rotation: a historical view. Astrophys Space Sci 328:269–277
18. Antia HK, Basu S (2001) Temporal variations in the solar rotation rate at high latitudes. Astrophys J 559:L69–L70
19. Schou J et al (1998) Helioseismic studies of differential rotation in the solar envelope by the Solar Oscillations Investigation using the Michelson Doppler Imager. Astrophys J 505:390–417
20. Foukal P, Lean J (1986) The influence of faculae on total solar irradiance and luminosity. Astrophys J 302:826–83. doi: 10.1086/164043
21. Le Mouël J-L, Shnirman MG, Blanter EM (2007)The 27-day signal in sunspot nand the solar dynamo. Solar Phys 246:295–307
22. Lean J L, Cook J, Maquette W, Johannesson A (1998) Magnetic sources of the solar irradiance cycle. Astrophys J 492:390–401
23. Dicke RH (1978) Is there a chronometer hidden deep inside the Sun. Nature 276:676–680
24. Vita-Finzi C (2010) The Dicke cycle: a 27 day solar oscillation. J Atmos Solar-Terr Phys 412(72):139–142
25. Hoyng P (1996) Is the solar cycle timed by a clock? Solar Phys 169:253–264
26. Lean J (1991) Variation in the Sun's radiative output. Rev Geophys 29:505–535
27. Foukal P, Bertello L, Livingston WC, Pevtsov AA, Singh J, Tlatov AG, Ulrich RK (2009) A century of solar Ca II measurements and their implication for solar UV driving of climate. Sol Phys 255:229–238
28. Floyd L, Tobiska K, Cebula RP (2002) Solar UV irradiance, its variation, and its relevance to the Earth. Adv Space Sci 29:1427–1440
29. Rottman G (2000) Variations of solar ultraviolet irradiance observed by the UARS SOLSTICE—1991 to 1999. Space Sci Rev 94:83–91
30. Thayer JP, Lei J, Forbes JM, Sutton EK, Nerem RS. 2008. Thermospheric density oscillations due to periodic solar wind high-speed streams. J Geophys Res 113:A06307. doi:10.1029/2008JA01319016]
31. Chandra S, McPeters RD (1994) The solar cycle variation of ozone in the stratosphere inferred from Nimbus 7 and NOAA 11 satellites. J Geophys Res 99:20665–20671
32. Hood LL (1999) Effects of short-term solar UV variability on the stratosphere. J Atmos Solar-Terr Phys 61:41–51

33. Gabis IP, Troshichev OA (2000) Influence of short-term changes in solar activity on baric field perturbations in the stratosphere and troposphere. J Atmos Solar Terr Phys 62:725–735
34. Wang Y, Cheng H, Edwards RL, He Y, Kong X, An Z, Wu J, Kelly MJ, Dykoski CA, Li X (2005) The Holocene Asian monsoon: links to solar changes and North Atlantic climate. Science 308:854–857
35. Giordano S, Mancuso S (2008) Coronal rotation at solar minimum from UV observations. Astrophys J 688:656–668
36. Parker GD (1976) Solar wind disturbances and recurrent modulation of galactic cosmic rays. J Geophys Res 81:3825–3833. doi:10.1029/JA081i022p03825
37. Takahashi Y, Okazaki Y, Sato M, Miyahara H, Sakanoi K, Hong PK (2009) 27-day variation in cloud amount and relationship to the solar cycle. Atmos Chem Phys Disc 9: 15327–15338
38. Tomczyk S, Schou J, Thompson M J (1995) Measurement of the rotation rate in the deep solar interior. Astrophys J Lett 448:L57–L60
39. Sturrock PA, Weber MA (2002) Comparative analysis of GALLEX-GNO solar neutrino data and SOHO/MDI helioseismology data: further evidence for rotational modulation of the solar neutrino flux. Astrophys J 565:1366–1375
40. Sturrock PA, Walther G, Wheatland MS (1997) Search for periodicities in the Homstake solar neutrino data. Astrophys J 491:409–413
41. Sturrock PA (2009) Combined analysis of solar neutrino and solar irradiance data: further evidence for variability of the solar neutrino flux and its implications concerning the solar core. Solar Phys 254:227–239
42. Pap J, Tobiska WK, Bouwer SD (1990) Periodicities of solar irradiance and solar activity indices. Solar Phys 129:165–189
43. Sturrock PA (2003) Time-series analysis of super-Kamiokande measurements of the solar neutrino flux. Astrophys J 594:1102–1107
44. Raaf JL (2008) Solar and atmospheric neutrinos in super-kaimokande. J Phys Conf Ser 136:022013. doi:10.1088/1742-6596/136/2/022013

Chapter 8
Contemporary History

Abstract Early attempts to monitor current changes in the Sun necessarily focused on sunspot patterns and solar diameter. The emphasis has shifted to the Sun's internal workings, its magnetism, and its output of radiation, plasma and particles, but thanks to improved instrumentation the direct assessment of changes in solar emission is now possible at intervals as short as a few microseconds and, besides revealing variations in the extent and timing of oscillations at different wavelengths, the new data make it possible to trace the evolution of individual sunspots, the onset and progress of flares, and changes in the magnetic flux. Global helioseismology yields information about solar structure in general while local helioseismology bears on three-dimensional structure and dynamics. Processes which underpin secular solar history, such as the creation of cosmogenic isotopes, are also clarified by the new data.

Keywords Helioseismology · SOHO · SDO · PICARD · Sunspot · Flare · TSI

Contemporary human history may be viewed as the period encompassed by living memory, although some accounts favour a definition which hinges on a technical advance such as nuclear technology or space flight. The equivalent in solar history could be the introduction of artificial satellites, but it would be shortsighted to disregard the intellectual context within which the observations are made, whereupon the acceptance of the Standard (p–p) Solar Model for nuclear fusion emerges as the defining criterion.

On the other hand a number of short-term regularities were (and still are) uncovered from ground-based data without benefit of orbiting devices or an agreed explanatory framework. Solar diameter, for example, has been studied for more than 350 years. Early measurements were used for determining the eccentricity of the Earth's orbit, which in turn fed into gravitational theory, and numerous estimates have been derived from observations of solar transits of Mercury and of solar eclipses. More recently astrolabe and sextant measurements have been supplemented by data derived from solar imaging.

C. Vita-Finzi, *Solar History*, SpringerBriefs in Astronomy,
DOI: 10.1007/978-94-007-4295-6_8, © The Author(s) 2013

Diameter

As noted in Chap. 1, the problem of defining a sphere when it is gaseous, compounded by atmospheric disturbance, hampers the use of early observations in analyses of secular change, and the data of helioseismology, though impressively precise, are model-based and bear on a diameter which is defined internally by reflection 4,000–8,000 km below the photosphere and is thus not readily compared with earlier, optical data [1].

Nevertheless there is every justification for investigating a record almost as long as that of sunspots. The measurements made by Picard and de la Hire in Paris during 1666–1719 have been taken to indicate decrease in solar diameter of about 4 arcsec [2]. Mercury transits during 1631–1973 pointed to changes which were apparently in phase with the Schwabe and Gleissberg cycles [3]. Others dispute these conclusions and claim, from analyses of data including these very transits of Mercury, that at least since 1715 there has been little or no discernible secular change [4].

Such a carefully worded verdict does not rule out oscillatory behaviour or indeed the possibility that a secular change will one day be discerned. Modern measurements under good observing conditions and with adequate instrumentation are beginning to yield convincing evidence of an association between radius and the solar cycle. Measurements with the Danjon astrolabe at Antalya, Rio de Janeiro and Santiago de Chile appear to show variations in solar radius in phase with magnetic activity [5]. Radius measurements from space using the michelson doppler instrument (MDI) on SOHO, once they had been corrected for temperature effects, also indicate a linear relationship between radius and sunspot number [6]. So has radio mapping of the solar disk at 48 GHz in 1991–1993 over half a solar cycle [7], while two flights of the Solar Disk Sextant (Fig. 8.1), in 1992 and 1994, indicated an increase in diameter by about 10 mas in two years.

There is some scepticism over the validity of these results too stemming from the techniques and technology used, discrepancies in the corrections adopted, and the lack of an agreed definition of solar diameter [3]. One of the key aims of the PICARD microsatellite programme (2010) is indeed to illuminate the links between solar variability and climate change by securing successive measurements of the photospheric solar diameter in the visible (535, 607 and 783 nm) and UV (230 nm) bands with an accuracy of 1 mas (1/1,000 arcsec). The project is also designed to identify active regions which might contaminate the diameter measurements.

Complementary observations in the south of France on a Danjon astrolabe and on a duplicate of the orbiting telescope should help to quantify the effect of atmospheric turbulence so that, although the PICARD results bear directly on a mere ~2 years, they will be complemented by subsequent, convincing ground-based measurements.

The diameter data for 1991–1993 at 48 GHz fluctuate with solar irradiance, as measured by the ERB and ACRIM II instruments, as well as sunspot number [7].

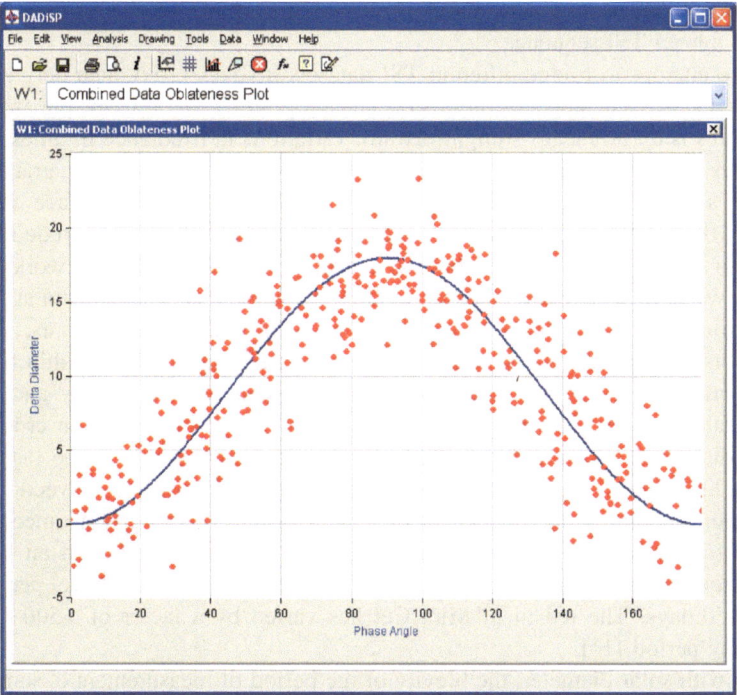

Fig. 8.1 The solar disk sextant (SDS) instrument was jointly developed by NASA and Yale University to measure the solar diameter at a very precise (milliarcs) level. The SDS has flown several times on stratospheric balloons. Each flight of the SDS generates four gigabytes of data (courtesy of DADiSP)

The crucial atmospheric level for the radio data corresponds to the base of the corona or the transition region, so that further confirmation of the link with solar dimensions would have a bearing on coronal heating. In other words the study of solar diameter has gained new significance in the study of processes as well as cyclic and progressive changes.

Total Solar Irradiance

In Chap. 6 we encountered the problems raised by the measurement gap between ACRIM I and ACRIM II and the possible secular change between Cycles 21 and 22. Trends in the short term are just as controversial. The ERB experiment included a cavity radiometer with a precision too low to detect intrinsic solar variability; the ACRIM I experiment (1980, 1984–1989) improved both accuracy and precision. Improvements continue to be made (e.g. [8]) but the outcome continues to provoke disputes especially in the context of the human contribution to climate change.

Instrument degradation and orbital effects further conspire to defeat attempts at a single agreed TSI sequence.

One attempt to bridge different TSI datasets invokes proxy data derived from the Sun's surface magnetic field [9]. The proposed link between the two solar attributes is by no means straightforward: variations in irradiance in timescales of minutes to hours appear to depend on solar oscillations and on granulation, whereas those on longer timescales are driven by changes in the surface magnetic field [10]. But this means that the irradiance record is more than a crude measure of solar activity and can contribute to analysis of the Sun's interior workings.

As we have seen, there is a longstanding dispute between those who argue that variations in TSI are wholly a product of surface magnetic features, such as sunspots, and those who ascribe them to structural readjustments within the Sun. The former could point to the association between surface activity and TSI at timescales measured in hours and months, as when an active region crosses the solar disk, and further argue that structural adjustments could not be faster than 1,000,0000 years, the thermal timescale for the base of the Sun's convection zone. But the structural response occurs on the timescale of the driving mechanism, namely the solar dynamo, and the response will occur in less than an hour. A convincing analogy is with Mira-type variable stars, which have periods of 150–450 days. The red giant Mira Cet has varied by a factor of >500 over its 332 day period [11].

As with solar diameter, the brevity of the period of measurement is sometimes offset by enhanced resolution. For instance there is some indication from the solar radiation and climate experiment (SORCE) spectral irradiance monitor (SIM) data that, at the last solar minimum (Cycle 23, AD 2006–2008), the infrared (IR at 1,600 nm) part of the irradiance was in phase with the TSI for short-term variation (such as the 27 day cycle) but not with the solar cycle (IR increased when the TSI decreased), prompting the suggestion that temperature changes were not uniform throughout the photosphere and perhaps even negative in deeper levels [12, 13]. But when it comes to explanatory hypotheses, such as the destabilization of active regions by the emergence of magnetic flux as the prerequisite for solar flares (Fig. 8.2), continuous, multiwavelength observation from space as well as from the ground still seems essential [14] especially as it largely avoids the complications of a fluctuating geomagnetic field.

The ultraviolet portion of the solar spectrum merits separate treatment because of its prominence in climatic analysis as well as its exaggerated variability over the Schwabe cycle. But any high resolution record has to be limited to the satellite era as the ozone layer shields the Earth more or less successfully from solar UV radiation. For instance the solar ultraviolet irradiance monitor (SUSIM) on the upper atmospheric research satellite (UARS) measured UV irradiance for wavelengths 115–410 nm daily at 1 and 5 nm resolution and weekly at 0.15 nm resolution between 1991 and 2005, that is to say in excess of a solar cycle. UARS also carried the solar stellar irradiance comparison experiment (SOLSTICE), a UV spectrometer which measured irradiance at 115–185, 170 and 320 nm daily during 1991–1999, that is over more than an average Schwabe cycle.

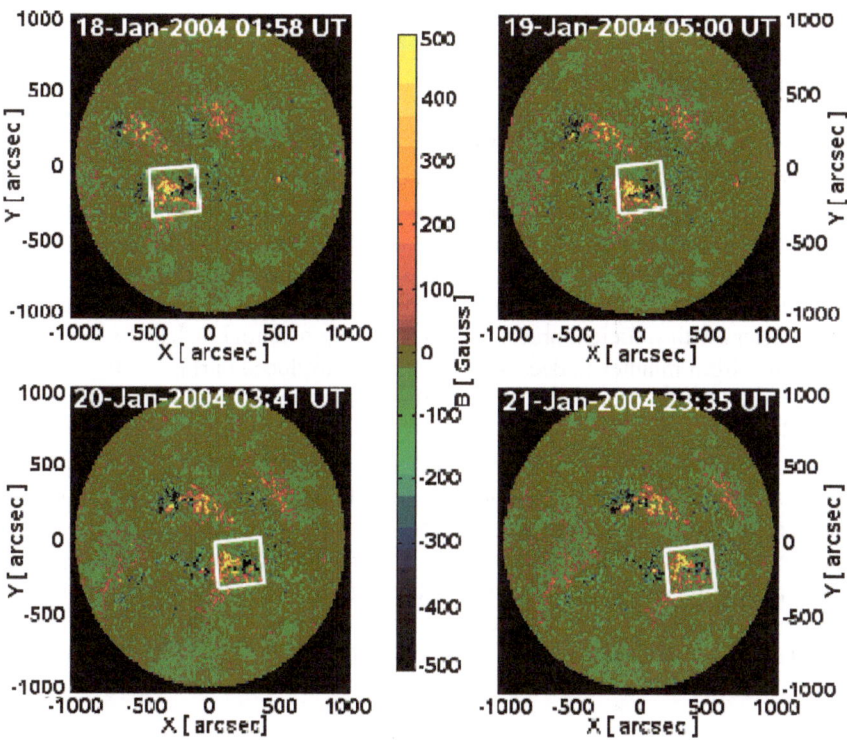

Fig. 8.2 SOHO/MDI LOS magnetograms for 18–21 January 2004 showing evolution of an active area (NOAA 10540) where magnetic field energy gradually rises until the release of an 6.1 flare releasing an estimated 10^{25} J. After [39] (courtesy of NASA)

Again, extreme ultraviolet levels in 2008 were 15 % lower than in 1996 but it is still unclear whether this is an instrumental rather than a solar trend [15]. In 1980 the order of magnitude of flux variation over the solar cycle of UV at 170–300 nm, responsible for photolysis of O_2 and O_3, was still unknown [16]. The EUV proxy $E_{10.7}$, as we saw in Chap. 6, has a potential time-resolution measured in minutes. The extreme UV imaging telescope (EIT) on SOHO imaged the corona at four UV wavelengths between 1996 and 2010; the EUV variability experiment on SDO can provide coverage between 0.1 and 105 nm continuously at a cadence of 10 s at 0.1 nm resolution and at a cadence of 0.25 s for six EUV bands [17].

The Transition Region and Coronal Explorer (TRACE), a space telescope in orbit during 1998–2010, obtained images at wavelengths ranging from visible light to far UV with an emphasis on the links between magnetic fields and plasma structures by observing the corona, the transition region and the photosphere with delays as little as 1 s and with a resolution of 1 arcsec. Its results have already proved the value of high resolution, high frequency imaging. For instance, EUV movies with a cadence of 5–10 s have revealed twisting in erupting coronal loops with fibrils measuring a few 1,000 km and associated with emerging ephemeral regions.

These advantages are enhanced with regard to the study of solar flares. Although they are observed in wavelengths ranging from gamma rays to radio frequencies emitted in all parts of the atmosphere, and occasionally in the photosphere, they are primarily driven by the dissipation of electric current in the corona [18]. Greatly improved understanding of the mechanism has come from detailed imaging and movies from space coupled with theoretical advances. For instance, Yohkoh (1991) revealed the existence of transient brightening lasting 1–10 min and of X-ray jets with velocities of 200 km s^{-1} in the corona [18]. Radiometers with sufficient precision and sampling potential to detect variations at timescales from minutes to decades were being introduced during 1980–1990 [19] and revealed 5 min fluctuations of ~ 0.003 % rising to 0.015 % during the largest solar flares [20].

The ability of TRACE to observe the Sun at temperatures ranging from 10^3 to 10^6 K allows it to record the buildup of magnetic energy in the corona as well as the reconnection that immediately precedes the main flare episode. Quasi-continuous observing benefits studies of the photosphere too where ground observation hinges on favourable weather and coordination between several observatories, witness the detailed analysis by TRACE in white light of a spot complex which produced three X-class flares on 6 and 7 June 2000 [21].

Sunspots

Sunspots became a device for objective solar studies once Galileo had demonstrated that they were part of the Sun and not planets or other bodies floating before it. Solar rotation, the preferred and changing location of sunspot groups, and the distinction between umbra and penumbra were soon identified. Heinrich Schwabe's discovery of the ~ 11 cycle came two centuries later, in 1843; the requisite data were collected by him rather than accruing over the intervening decades. From the outset, telescopic observation of sunspots was seen to demonstrate variations in their morphology as well as in their number and distribution. The Zurich classification of sunspot groups has an evolutionary basis [22].

The first physical investigation into an individual sunspot was by Alexander Wilson in 1769 when he reported observations suggesting that spots represent depressions in the Sun's surface. The next key step was taken by Ellery Hale in 1908 when he launched the study of sunspot magnetic fields [22].

The intervening centuries have witnessed advances in the categorisation of sunspot morphology underpinned by improved understanding of the underlying processes and, thanks to new techniques, in the resolution with which flows and oscillations can be traced. Thus by 2004 it was possible to resolve features with a width of 70 k. The advances evidently owed much to observation from space, notably by TRACE but ground-based observation has also made significant progress thanks to the development of adaptive optics, the technique whereby a deformable mirror reshaped up to 130 times/second compensates for atmospheric blurring. The Dunn telescope in New Mexico can obtain 0.14 arcsec resolution rather than the 0.5–1.0 arcsec that would otherwise be the limit for observation from the ground.

This is not to ignore the improved appreciation of sunspot history, as well as related features of the photosphere, arising from instrumental improvements coupled with advances in photography. The findings bear on the fine structure of sunspots, granulation in the penumbra, and the evolution of sunspot groups. Video imaging and stereoscopy have further demonstrated the need to combine studies of individual sunspot properties with statistical analysis of sunspot occurrence and distribution.

Prior to the satellite era, the fine detail of a single sunspot was considered longlived when compared to photospheric granulation: some 2 h against ~ 10 min. Sunspots were thought to originate in pores, that is spots with diameters of 1–5 arcsec and occasionally as much as 10 arcsec and lacking a penumbra. The transition from pore to spot was marked by a rudimentary penumbra; fully-fledged penumbrae had a filamentary structure. The strength and orientation of the sunspot's magnetic field could be measured by the Zeeman effect whereby, when a spectroscope scans across a magnetically active area, a single spectral line is split into two or three strands, the width of the field being proportional to the field strength. But the exact morphology of the field was not reliably known [22].

The spot itself was long thought to indicate inhibition of the flow of heat to the surface by enhanced magnetic activity. A recent suggestion is that the cooler conditions allow hydrogen atoms to bond into H_2 molecules, leading to reduced pressure and thus further intensification of the magnetic field [23]. The H_2 was detected by its proxy OH, whose spectral line (at 1565 nm) is fortuitously located near that for Fe I (at 1564.8 nm) for which the authors' spectrograph was designed. The theory does not explain the initial magnetic focusing but it accounts for its intensification.

The resolution of still images of sunspots increased from 0.2 arcs to 0.12 in 2002 once the 1 m solar telescope on La Palma, which uses adaptive optics, came into operation. It revealed, among other things, that penumbral filaments contain dark cores less than 90 km across. In May 2006 two simultaneous time-series taken by TRACE in white light and in UV (170 nm) of a small sunspot (NOAA 10886) showed horizontal, radial outflow to a distance of four times the radius of the spot, and there was evidence of inward motion in the penumbra.

The TRACE pixel size (0.5 arcsec) was too coarse for detailed analysis of the related fine structures [24]. On the other hand computer simulation, made possible by advances in supercomputing, provides insights into the dynamics at issue and in particular the changes in magnetic field between umbra and penumbra (Fig. 8.3). And Doppler measurements by MDI allow deeper sunspot structure to be imaged to a depth of 22,000 km (as in the example from June 1998 in Fig. 8.4).

To be sure, the image is couched in sound speed, using the travel-time of solar surface waves (f-modes), but that translates into temperature: gas cooled by the

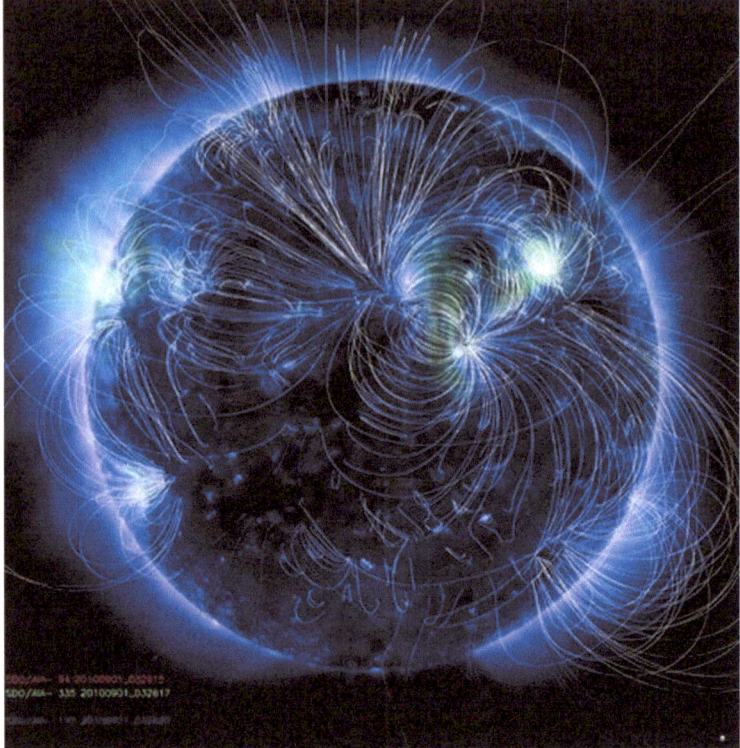

Fig. 8.3 Solar atmosphere in three wavelengths imaged by the Atmospheric Imaging Assembly (AIA) instrument on SDO with modelled field lines. The AIA images the Sun in 10 wavelengths every 10 sec (image courtesy SDO/NASA)

sunspot's magnetic field flows down to be replaced by more gas. This imports its own magnetic field and thus prevents dissipation of the strong sunspot field, so that the system is self sustaining [25].The sunspot is a shallow phenomenon measuring 5,000–6,000 km, and the observations provide the first direct evidence for the model proposed by Parker [26] in which a loose cluster of magnetic flux tubes is held together by downflow beneath the sunspot.

The Magnetic Flux

Many aspects of solar magnetism call for a combination of long-term and short-term data, the former (measured in years) to establish trends and any association with the solar cycle, and the latter to document processes in near real time. The resulting benefits are illustrated by the improved understanding of the field reversal that characterizes the photospheric magnetic field every ~ 11 years.

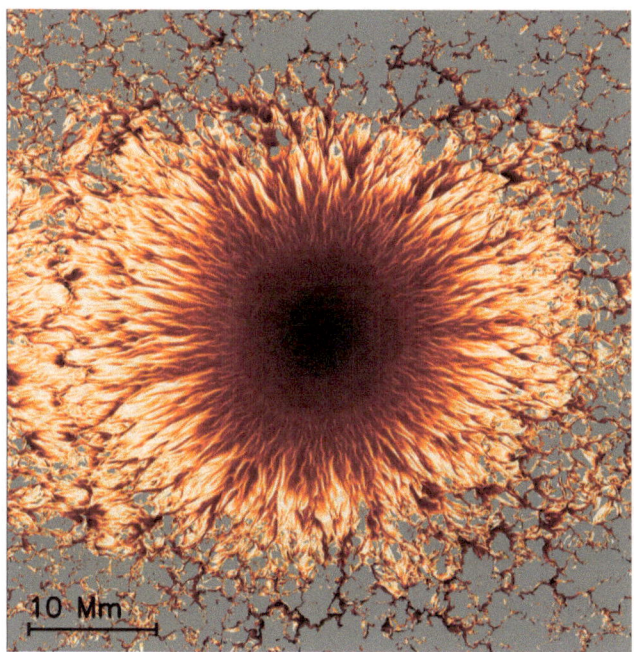

Fig. 8.4 Result of computer simulation of sunspot dynamics showing narrow, almost horizontal (lighter to white) filaments embedded in more vertical (darker to black) magnetic field. 10 Mm = 10,000 km; grey: field strength < 200 G. (©UCAR, image courtesy Matthias Rempel, NCAR with permission.)

Satellite magnetometers at 1 AU have demonstrated that the reversal is not confined to the familiar pattern of sunspot migration in the photosphere but extends through the corona into the heliosphere. Magnetic reconnection, the reconfiguration of magnetic fields and coronal mass ejections play a part in the essential task of removing magnetic flux from the photosphere and from the corona into space. In short, the solar atmosphere (Fig. 8.5) is truly coupled to the solar dynamo [27].

The disturbance undergone by the Earth's magnetic field in response to solar events, epitomized by the Carrington event of 1859 [28], also requires this two-pronged approach. It benefits from the existence of a global measure of the disturbance, the aa index, mentioned in Chap. 6. There is a close link to sunspot number. The solar wind fills the heliosphere with magnetic field, and magnetic reconnection allows energy from the solar wind to penetrate the near-Earth interplanetary magnetic field (IMF) [29]. According to NOAA the general aa level has increased 'substantially' since 1900. Using data from solar cycles 20–22 it was found that the coronal source flux had increased by a factor of 1.4 since 1964 and, to judge from surrogate measurements, by a factor of 2.3 since 1901 [29, 30].

The assumption that the aa index bears a constant relationship to the IMF has been challenged [31]. Equally, even if the interplanetary field strength has

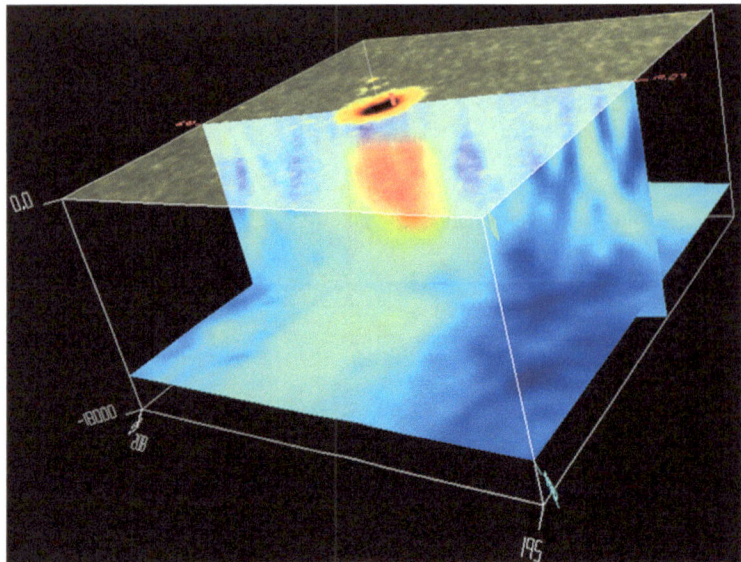

Fig. 8.5 Subsurface structure below a sunspot based on sound speed derived from Doppler measurements by MDI. Areas of faster sound speed in reds, slower sound speed in blues. Bottom plane at 22000 km shows horizontal variation of sound speed (courtesy of NASA/GSFC/SOHO)

doubled, it does not follow that the total flux has undergone a similar increase [32]. There are obvious implications in these objections for the use of surrogate sources, in the reconstruction of solar history, as the open flux, which modulates the cosmic ray flux at Earth, is rooted in coronal holes, and the total photospheric field, rooted in plages and sunspots, strongly influences the solar radiative output.

How far does all this invalidate the use of [10]Be as portmanteau measure of past solar activity? There is a prima facie case for it, at the intermediate timescale of the instrumental record, in the anticorrelation of solar EUV radiation with cosmic ray receipts and [10]Be flux, but the links in the causal chain are loose. Fortunately the already admirable aa record is being daily enriched by data coordinated at ESA's Redu Station in the Ardennes, in Belgium, and by formal assessment of geomagnetic storms.

Geomagnetic storms are now defined [33] by the Kyoto Dst index, which represents the axially symmetric disturbance magnetic field at the dipole equator on the Earth's surface. It uses data from four high quality observatories (Kakioka, Japan; Hermanus, South Africa; San Juan, Puerto Rico; and Honolulu, Hawai'i) which are located far from the equatorial and auroral electrojets and widely separated in longitude. Positive variations are caused mainly by compression of the magnetosphere by solar wind pressure.

Since 1957 the Kyoto Dst index has been determined hourly. Its coverage has been criticized as insufficient and asymmetrical [34] but it yields valuable results. For instance it revealed a cumulative change in solar wind speed from 303 km/s in

1902 to 545 km/s in 2003 [35]. But for shortlived events the resolution is inadequate, whence the development of the Kyoto Sym-H, using data from mid-latitude observatories, with a resolution of 1 min, and an equivalent index for the low-latitude Dst observatories for 1985–2007 [36]. The AL index is based on the horizontal component of the geomagnetic field along the auroral oval in the Northern Hemisphere (above 50° N) and is compiled every minute by selecting the most negative values among 12 stations [37]. The Dst index is a guide to storm activity, the AL index to storm intensity.

The real-time solar wind magnetosphere-ionosphere system (WINDMI), a nonlinear dynamic model of the coupled magnetosphere-ionosphere system, has been used to predict Dst and AL values about 1 h before the onset of geomagnetic substorms and storms with a threshold of -50 nT for the predicted Dst and -400 nT for the predicted AL. Input solar wind driving voltages are derived from real-time measurements of solar wind proton density, solar wind velocity and IMF from the ACE spacecraft at 1 min intervals.

The results [38] are encouraging as regards rapid forecasting of magnetic storms, essential for safeguarding communication, global positioning and other spacecraft, and for testing models on the basis of short-term observational history. The next challenge will be to identify how far the geomagnetic field undergoes solar distortion when solar flares and other disturbances are not manifested, that is to say when processes within the Earth dominate.

References

1. Dziembowski WA, Goode PR, di Mauro MP, Kosovichev G, Schou J (1998) Solar cycle onset seen in SOHO Michelson Doppler imager seismic data. Astrophys J 509:456–460
2. Ribes E, Ribes JC, Barthalot R (1987) Evidence for a larger Sun with a slower rotation during the seventeenth century. Nature 326:52–55
3. Thuillier G, Sofia S, Harberreiter M (2005) Past, present and future measurements of the solar diameter. Adv Space Sci 35:329–340
4. Parkinson J H, Morrison LV, Stephenson F R (1980) The constancy of the solar diameter over the past 250 years. Nature 288:548–551
5. Noël F (2002) On solar radius measurements with Danjon astrolabes. Astr Astrophys 396:667–672. doi: 10.1051/0004-6361:20021447
6. Emilio M, Kuhn JR, Bush RI, Scherrer P (2000) On the constancy of the solar diameter. Astrophys J 543:1007–1010
7. Costa JER, Silva AVR, Makhmutov VS, Rolli E, Kaufmann P, Magun A (1999) Solar radius variations at 48 Ghz correlated with solar irradiance. Astrophys J 520:L63. doi: 10.1086/312132
8. Kopp G and Lean JL (2011) A new, lower value of total solar irradiance: evidence and climate significance. Geophys Res Lett 38:L017106. doi: 10.1029/2010GL045777
9. Scafetta N, Willson RC (2009) ACRIM-gap and TSI trend issue resolved using a surface magnetic flux TSI proxy model. Geophys Res Lett 36:L05701. doi: 10.1029/2008GL036307
10. Krivova NA and Solanki SK (2008) Models of solar irradiance variations: current status. J Astrophys Astron 29:151–158
11. Sofia S (2004) Variations of total solar irradiance produced by structural changes of the solar interior. EOS Trans AGU 85:217, 221

12. Harder JW, Fontenla JM, Pilewskie P, Richard EC, Woods TN (2009) Trends in solar spectral irradiance variability in the visible and infrared. Geophys Res Lett 36:L07801. doi: 10.1029/2008GL036797
13. Woods TN (2010a) Irradiance variations during this solar cycle minimum. In: Cranmer S, Hoeksema T, Kohl J (eds) SOHO-23: understanding a peculiar solar minimum. ASP Conf Ser 428:68
14. Oliver R, Ballester JL, Baudin F (1998) Emergence of magnetic flux on the Sun as the cause of a 158 day periodicity in sunspot areas. Nature 394:552–553
15. Woods TN (2010b) SORCE science meting session 3/3.06. At lasp.colorado.edu/sorce/news
16. Foukal P (1980) Solar luminosity variation on short time scales; observational evidence and basic mechanisms. In: Pepin RO, Eddy JA, Merrill RB (ed) The ancient Sun. Pergamon, New York
17. Woods TN and 13 others (2011) New solar extreme-ultraviolet irradiance observations during flares. Astrophys J 739:59. doi: 10.1088/0004-637X/739/2/59
18. Shibata K and Magara T (2011) Solar flares: magnetohydrodynamic processes. Living Rev Solar Phys 8:1–99. Online at www.livingreviews.org/lrsp-2011-6
19. Fröhlich C, Foukal PV, Hickey JR, Hudson HS, Willson RC (1980) Solar irradiance variability from modern measurements. In: Sonett CP, Giampapa MS, Matthews MS (eds) The Sun in time, University of Arizona, Tucson
20. Fröhlich C and Lean J (2004) Solar radiative output and its variability: evidence and mechanisms. Astron Astrophys Rev 12:273–320
21. Schrijver (2001) 2001 senior review proposal for *TRACE*. Lockheed martin missiles and space/ATC
22. Bray RJ, Loughhead RE (1964) Sunspots. Chapman and Hall, London
23. Jaeggli SA, Lin H, Uitenbroek H (2012) On molecular hydrogen formation and the magnetohydrostatic equilibrium of sunspots. Astrophys J 745:13. doi: 10.1088/0004-637X/745/2/133
24. Balthasar H, Muglach K (2010) The three-dimensional structure of sunspots. II. The moat flow at two different heights. Astr Astrophys 511:A67. doi: 10.1051/0004-6361/200912978
25. Zhao J, Kosovichev AG, Duvall TL Jr (2001) Investigation of mass flows beneath a sunspot by time-distance helioseismology. Astrophys J 557:384–388
26. Parker EN (1979) Sunspots and the physics of magnetic flux tubes. I. The general nature of the sunspot. Astrophys J 230:905–913
27. Low BC, Zhang M (2003) Global magnetic-field reversal in the corona. In: Pap JM, Fox P (eds) Solar variability and its effects on climate, AGU, Washington
28. Tsurutani BT, Gonzalez WD, Lakhina GS, Alex S (2003) The extreme magnetic storm of 1–2 September 1859. J Geophys Res 108:1268. doi: 1029/2002JA009504
29. Lockwood M, Stamper R (1999) Long-term drift of the coronal source magnetic flux and the total solar irradiance. Geophys Res Lett 26:2461–2464
30. Lockwood M, Stamper R, Wild MN (1999) A doubling of the Sun's coronal magnetic field during the last 100 years. Nature 388:437–439
31. Svalgaard L, Cliver EW, Le Sager P (2002) No doubling of the Sun's coronal magnetic field during the last 100 years. At www.leif.org/research/No%20Doubling.pdf
32. Wang Y-M, Lean J, Sheeley NR Jr (2002) Role of a variable meridional flow in the secular evolution of the Sun's polar fields and open flux. Astrophys J 577:L53–L57
33. Gonzalez WD, Joselyn JA, Kamide Y, Kroehl HW, Rostoker G, Tsurutani BT, Vasyliunas VM (1994),. What is a geomagnetic storm? J Geophys Res 99:5771–5792
34. Karinen A, Mursula K, Takalo J, Ulich Th (2002) An erroneous Dst index in 1971. ESA SP-4787, Vico
35. Svalgaard L, Cliver EW (2007) Interhourly variability index of geomagnetic activity and its use in deriving the long-term variation of solar wind speed. J Geophys Res 112:A10111. doi: 10.1029/410 2007JA012437
36. Gannon JL, Love JJ (2010) USGS 1-min Dst index. J Atmos Sol-Terr Phys. doi : 10.1016/j.jastp.2010.02.013

37. Mays ML, Horton W, Spencer E, Kozyra J (2009) Real-time predictions of geomagnetic storms and substorms: use of the solar wind magnetosphere-ionosphere system model. Space weather 7:S07001. doi: 10.1029/2008SW000459
38. Newell PT, Sotirelis T, Liou K, Rich FJ (2007) A nearly universal solar wind-magnetosphere coupling function inferred from 10 magnetospheric state variables. J Geophys Res 112:A01206. doi: 10.1029/2006JA012015
39. Thalmann JK, Wiegelmann T (2008) Evolution of the flaring active region NOAA 10540 as a sequence of nonlinear force-free field extrapolations. Astron Astrophys 484:495–502. doi: 10.1051/0004-6361:200809508

Chapter 9
Lessons from History

Abstract The long-term history of the Sun is patchy, being based mainly on the implications of the Standard Solar Model and the record of flares recovered from the lunar regolith and meteorites exposed to cosmic rays at different times. However the fossil neutrino evidence, especially from tellurides, may eventually provide successive snapshots of the solar interior. Cyclical and aperiodic changes in TSI and UV are well documented for recent millennia, although they may turn out to be minor oscillations with wavelengths measured in $\geq 10^4$ yr, and they provide the basis for approximate but robust forecasts for solar behaviour and its impact on space weather and terrestrial climate.

Keywords Standard Solar Model · Flare · Neutrino · TSI · UV · Space weather · Climate · Forecast · Heliogeology

Lessons from history does not mean the same as the lessons of history—morality audits that seem generally to disillusion their authors—but neutral accumulations of useful data structured by chronology. *Solar history* aims simply to improve understanding of the Sun's behaviour by tracing its evolution from its early, murky beginnings to its exhaustively documented present. There are numerous accounts of one or other but so far as one could tell none that joined them up. At the very least such an account would allow long-wavelength periodicities to emerge (or prove to be absent), cumulative changes to be quantified, and the Sun to illuminate our understanding of the 100 billion other Sun-like stars in our home galaxy and thus of their Earth-like acolytes.

There was also the hope of improving current solar models and launching new ones in the light of novel data or unfamiliar questions. Regrettably, many items which bear on space weather, climate change and resource management have become politicized or ethically charged, to the detriment of their detached analysis even if there are potential benefits to be gained from the enhanced resources that are in consequence devoted to data acquisition. In any case, prediction would be aimed mainly at evaluating explanations, with any harvests and satellites saved on the way as a bonus.

C. Vita-Finzi, *Solar History*, SpringerBriefs in Astronomy,
DOI: 10.1007/978-94-007-4295-6_9, © The Author(s) 2013

Cumulative Change

Solar history as sketched here extends from 10^9 to $<10^{-9}$ year, from the aeons of the early solar system to the fractions of a second required for an instrument on a satellite to capture a scene or burst of radiation.

The progress of only one component can be followed throughout: the Sun's ageing as deduced from analogy with other stars of similar mass. The progressive conversion of hydrogen to helium is of course likely to be complicated by changes in the structure of the Sun's convective zone, some of them characterized by timescales as short as months [1]. The estimated rate of 6. 10^8 tonnes/s for the conversion of hydrogen to helium is of course modulated by associated temperature changes but, assuming hydrostatic equilibrium is maintained, the resulting core contraction is offset by increased luminosity.

Thus one must assume that the faint young Sun, which attained ~ 140 % of its luminosity at birth, continues to brighten. But the process cannot be documented by satellite data: as we saw in Chap. 6, the ACRIM I-II record can be taken to indicate an increase in total irradiance of ~ 0.03 % between Cycles 21 and 22, but the equivalent ~ 3 %/1100 yr seems extravagant. Even the ~ 30 kyr of decline in cosmic ray flux indicated by ^{10}Be and ^{14}C ice core data (Chaps. 4 and 5; see also [2]) cannot provide a helpful secular measure of irradiance change because they represent the effect of the solar wind and geomagnetism on the cosmic ray flux rather than the solar component alone and they may fall within gross solar maxima and minima which deviate from the general trend. The best hope for monitoring the SSM resides in mining fossil neutrinos from tellurides of different age (Chap. 1) and thus in drastic reduction of the cost and labour required for their analysis.

The only element of the history that can be traced throughout the narrative is solar flare activity, and it bears primarily on the corona and thus only very indirectly on the solar interior. Nevertheless the flare record, though it is at risk of losing the ice-core (NO_x) evidence, provides insights into the Sun's behavior immediately before it joined the Main Sequence, when rapid rotation and a storehouse of ionized gases liberated the violent flares that qualify the protoSun as a T Tauri star. The evidence will gain in value from current satellite observations fed into the empirical Flare Irradiance Spectral Model (FISM), which extends from 0.1 to 193 nm, by quantifying the contribution by flares to different parts of the solar spectrum in the vacuum (< 200 nm) UV range and quantifying their potential photochemical and thermal effects.

The longest flare record of ~ 1 Gyr is currently held by the Moon but it is marred by a stratigraphy drastically limited by a thin regolith which has been greatly disturbed. Some evidence of changes in flare incidence have been detected, including a possible surge by a factor of 20–40 about 20 ± 10 kyr ago [3], but nothing amounting to a cumulative change has so far been claimed. The meteorite evidence is evidently highly discontinuous, as it hinges on the arrival and discovery of suitable bodies, but the chronology is potentially clearer as it is based mainly on separate specimens rather than a blurred stratigraphy.

Fluctuations in solar activity, whether directly observed or derived from cosmogenic isotopes or other indirect sources, embody cycles which range from ~ 11 to ~ 2300 years. Longer wavelengths are not ruled out. The significance for the physics of the solar interior remains obscure. The 27-day irradiance cycle is generally ascribed to the passage of active areas across the solar disk consequent on its rotation but there are indications that instead it is a response to rotation of the Sun's radiative zone.

In short, empirical solar history lags behind the inferences of the modellers, perhaps because it is targeted at the needs of meteorology more than of stellar astronomy and cleaves to uniformitarianism [4] when geoscience now views this as an unnecessary straitjacket. But great advances in meteoritics at one extreme and satellite observation at the other should soon reduce the uncertainty.

Forecasting

Solar activity, it is claimed [5], can be usefully predicted at different timescales: one that is measured in hours or rotation periods, with related flare effects; the 11/22 year cycle within the Gleissberg (~ 88 year) envelope and complicated by the Maunder effect (see also [6]); and an unspecified time frame for forecasting flares and CMEs from the evolution of active regions and the detection of loop arcades in X-rays and the like from satellites. In the light of preceding chapters one can modify the scheme to distinguish between forecasts, which describe future solar behavior in broad terms, as in the circumstances that commonly accompany solar maxima and minima, and predictions, which build on sufficient observational data for a high degree of certainty, as in the likely development of a CME.

Forecasts of solar cycles employ statistical, spectral and precursor methods and may invoke planetary influence. In 1997 a NOAA-led panel urged the scientific community to develop a fundamental understanding of the solar cycle that would provide a physical rather than an empirical basis for prediction methods [7]. Yet the two approaches are complementary: for instance, the delineation of an incoming solar cycle (Fig. 9.1) may hinge on such empirical details as the centroid position of sunspot areas for each solar rotation since 1874 analyzed as functions of time from each sunspot cycle minimum [8], but the results help us to discriminate between different explanatory models [9].

Even so, the wide range of estimates that have been made for the timing and magnitude of Cycle 24 demonstrates that true prediction here still remains problematic. Forecasts of this or other features of solar behaviour will thus continue to build on the improved circumstantial evidence that comes from solar cycle analysis of cosmogenic isotopes and assessment of TSI variation during grand maxima and minima. They would also exploit the insights into millennial UV variation deriving from palaeoclimate and palaeohydrology. In other words, as in terrestrial climatology, the strategy must be to broaden both the temporal reach and the range of data sources to enrich the basis of the forecasts. A long history is a prerequisite.

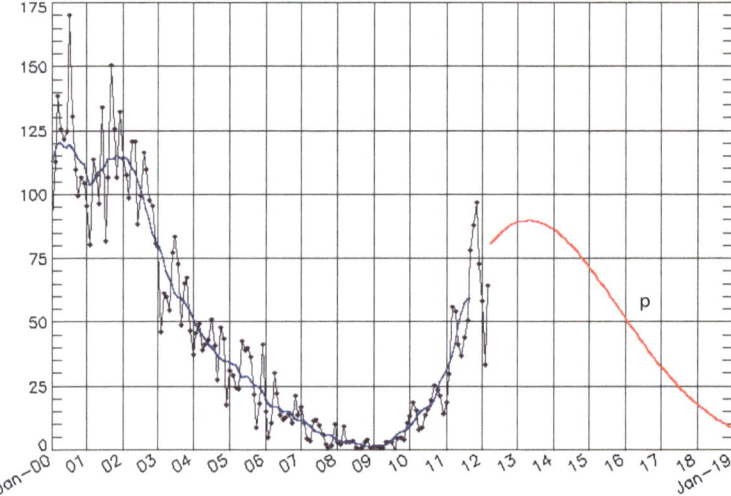

Fig. 9.1 Forecasting the next 11-year cycle on the basis of data observed until March 2012. *Dots*: monthly sunspot number; curves show smoothed sunspot number and p: predicted values (courtesy of NOAA/SWPC Boulder, CO)

In contrast, as the specificity, precision and time-resolution of direct observation improves, so is prediction likely to become narrower in focus though still founded on analysis of antecedents. The Advance Composition Explorer (ACE), for example, secures solar wind data of such frequency that it can give up to an hour's warning of unusual solar behaviour and thus, as a bonus, forecast CMEs and solar flares (Fig. 9.2). The Atmospheric Imaging Assembly (AIA) on the SDO observes on 7 channels every 12 s; on 13 June 2010 it targeted a flare with a life of 10 min and was able to document its rise, the energy release, and the changing temperature that characterized the early stages of the resulting CME [10]. Naturally the great increase in the flow of data requires a corresponding statistical armoury (e.g. [11]) and increasing reliance on automated space weather alerts [12].

Heliogeology

One can foresee the blossoming of a research field which does not simply combine the data of geosciences and of solar astrophysics but targets those areas where a combined effort seems justified.

There are two helpful precedents in naming. The first is astrogeology, which in fact is commonly confined to planetary and meteoritic rather than stellar research. The second is heliobiology, a branch of biophysics with Svante Arrhenius among its early practitioners which investigates the effect of changes in solar activity on

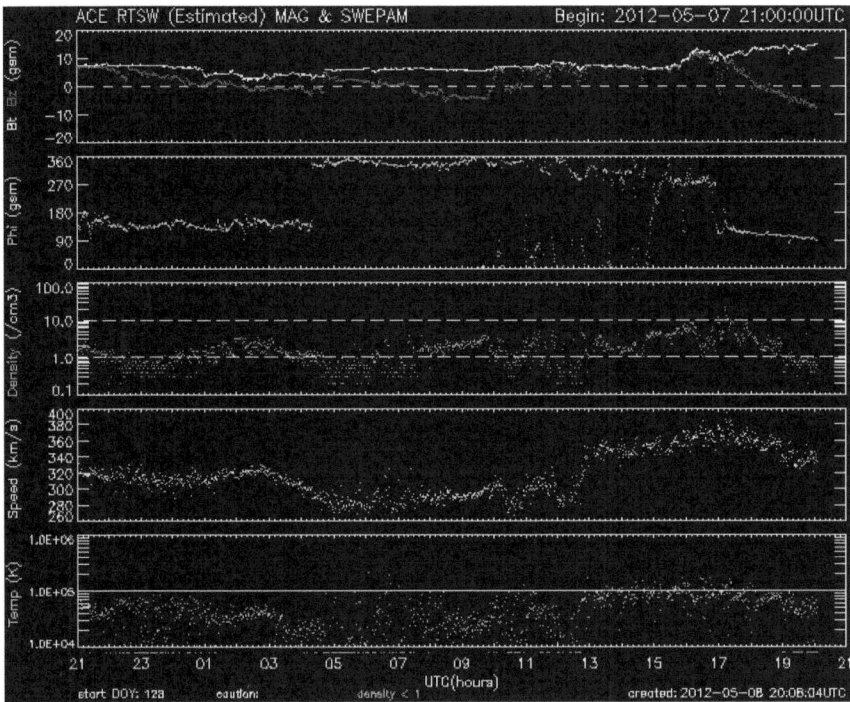

Fig. 9.2 Advance Composition Explorer (ACE) real time data including solar wind (RTSW) temperature and speed (courtesy of NOAA)

terrestrial organisms, including familiar items such as the growth of tree rings as well as more esoteric themes including epidemics, locust plagues and the pathogenicity of certain microorganisms [13]. The hybrid name *heliogeology* seems appropriate for this borderland.

References

1. Dearborn DS (1991) Standard solar models. In: Sonett CP, Giampapa MS, Matthews MS (eds) The Sun in time. University Arizona, Tucson AZ
2. Steinhilber F, Beer J, Fröhlich C (2009) Total solar irradiance during the Holocene. Geophys Res Lett 36:L19704. doi:10.1029/2009GL040142
3. Zook AH (1980) On lunar evidence for a possible large increase in solar flare activity approximately 2×10^4 years ago. In: Pepin RO, Eddy JA, Merrill, RB (eds) The ancient Sun. Pergamon, New York
4. Eddy JA (1988) Variability of the present and ancient Sun: a test of solar uniformitarianism. In: Stephenson FR, Wolfendale AW (eds) Secular solar and geomagnetic variations in the last 10,000 years. Kluwer, London
5. Foukal PV (2004) Solar astrophysics, 2nd edn.Wiley, Berlin

6. Tobias SM, Weiss Beer J (2004) Long-term prediction of solar activity. Astron Geophys 45, 2.06
7. Joselyn JA, Anderson JB, Coffey H, Harvey K, Hathaway D, Heckman G, Hildner E, Mende W, Schatten K, Thompson R, Thompson AWP, White OR (1997) Panel achieves consensus prediction of solar cycle 23. EOS 78:205
8. Hathaway DH (2011) A standard law for the equatorward drift of the sunspot zones. Solar Phys 273:221–230. doi: 10.1007/s11207-011-9837-z
9. Rempel M, Schlichenmaier R (2011) Sunspot modelling: from simplified models to radiative MHD. Living Rev Solar Phys 8:3
10. Patsourakos S, Vourlidas A, Stenborg G (2010) The genesis of an impulsive CME observed by AIA on SDO. AGU Fall Mtg Abs #SH14A-03
11. Reale F, Ciaravella A (2006) Analysis of a multi-wavelength time-resolved observation of a coronal loop. Astron Astrophys 449:1177
12. Lobzin VV, Cairns IH, Robinson PA, Steward G, Patterson G (2010) Automatic recognition of coronal type II radio bursts: the automated radio burst identification system method and first observations. Astrophys J Lett 710:L58. doi: 10.1088/2041-8205/710/1/L58
13. Platonova AT (1979) Heliobiology. Great Soviet Encyclopedia, 3rd edn. Gale, Farmington Hills MI

Index

C. Vita-Finzi, *Solar History*, SpringerBriefs in Astronomy,
DOI: 10.1007/978-94-007-4295-6, © The Author(s) 2013